REVISED
STUDENT SOLUTIONS MANUAL

NINTH EDITION
MODERN ELEMENTARY
STATISTICS

REVISED
STUDENT SOLUTIONS MANUAL

JOHN E. FREUND · GARY A. SIMON
ARIZONA STATE UNIVERSITY   NEW YORK UNIVERSITY

NINTH EDITION
MODERN ELEMENTARY
STATISTICS

PRENTICE HALL, Upper Saddle River, New Jersey 07458

Production Editor: ***Kimberly Dellas***
Production Coordinator: ***Alan Fischer***
Acquisitions Editor: ***Millicent Treloar***
Special Projects Manager: ***Barbara A. Murray***
Cover Designer: ***PM Workshop Inc.***
Supplement Cover Manager: ***Paul Gourhan***

 © 1997 by **PRENTICE-HALL, INC.**
Simon & Schuster/A Viacom Company
Upper Saddle River, NJ 07458

All rights reserved. No part of this book may be reproduced, in any form or by any means,
without permission in writing from the publisher.

Printed in the United States of America

10 9 8 7 6 5 4 3 2 1

ISBN 0-13-858317-X

Prentice-Hall International (UK) Limited, *London*
Prentice-Hall of Australia Pty. Limited, *Sydney*
Prentice-Hall Canada, Inc., *Toronto*
Prentice-Hall Hispanoamericana, S.A., *Mexico*
Prentice-Hall of India Private Limited, *New Delhi*
Prentice-Hall of Japan, Inc., *Tokyo*
Simon & Schuster Asia Pte. Ltd., *Singapore*
Editora Prentice-Hall do Brasil, Ltda., *Rio de Janeiro*

TABLE OF CONTENTS

		Page
Chapter 1	Introduction	1
Chapter 2	Summarizing Data: Frequency Distribution	3
Chapter 3	Summarizing Data: Measures of Location	9
Chapter 4	Summarizing Data: Measures of Variation	15
Chapter 5	Possibilities and Probabilities	21
Chapter 6	Some Rules of Probability	27
Chapter 7	Expectations and Decisions	37
Chapter 8	Probability Distributions	41
Chapter 9	The Normal Distribution	51
Chapter 10	Sampling and Sampling Distributions	55
Chapter 11	Inferences About Means	63
Chapter 12	Inferences About Standard Deviations	73
Chapter 13	Inferences About Proportions	77
Chapter 14	Analysis of Variance	91
Chapter 15	Regression	99
Chapter 16	Correlation	105
Chapter 17	Nonparametric	111

CHAPTER 1

Introduction

1.1 The following are possibilities:
   (a) In a random sample of 200 women under thirty, 137 stated that they prefer perfume as a gift to either flowers or candy. At the 0.05 level of significance, does this refute the claim that 60 percent of all women under thirty prefer perfume as a gift to either flowers or candy?
   (b) In a random sample of 200 violinists, 137 stated that they started playing the violin before they were ten years old. At the 0.05 level of significance, does this refute the claim that 60 percent of all violinists start playing the violin before they were ten years old?
   (c) In a random sample of 200 cars coming to a certain intersection, 137 made right turns. At the 0.05 level of significance, does this refute the claim that 60 percent of all cars coming to this intersection make right turns?

1.3 (a) Persons coming out of the building housing the national headquarters of a political party are more likely to support that party.
   (b) December (pre-Christmas) spending is not typical of spending throughout the year.

1.5 The following are possibilities:
   (a) An experiment was done to compare two fertilizers. Fertilizer A improved production in 14 of 45 fields while fertilizer B improved production in 8 of 41 fields. At the 0.05 level of significance, does this information refute the claim that the two fertilizers are equally effective?
   (b) An experiment was done to compare two groups of voters. In group A, 14 of 45 approved of Mr. Clinton, while in group B, 8 of 41 approved of Mr. Clinton. At the 0.05 level of significance, does this information refute the claim that the two groups have the same opinion about Mr. Clinton?
   (c) An experiment was done to compare two instructional methods. With method A, 14 of 45 cases reported showed effectiveness, while with method B, 8 of 41 cases reported showed effectiveness. At the 0.05 level of significance, does this information refute the claim that the two methods are equally effective?

1.7 (a) Since only 79 and 88 exceed 75, the conclusion follows from the data.
   (b) If we assume that the student received the grades in the given order, the conclusion follows from the data.
   (c) Since there may have been other reasons, the conclusion goes beyond the data.
   (d) Since 88 − 46 = 42, the conclusion follows from the data.

1.9 (a) and (b) are numeric descriptions of the data, while (c) and (d) make generalizations which go beyond the data.

1.11 (a) ordinal; (b) nominal; (c) ordinal; (d) nominal. In (d), removal of the "no opinion" category would leave ordinal data.

1.13 If IQ scores are interval data, then the IQ difference between 135 and 105 is three times the difference between 105 and 95. Since IQ scores are rescaled z-scores (see section 4.3), it is unrealistic to interpret the above as three times a "smartness" difference.

1.15 (a) Since there may have been other reasons for the low attendance, the conclusion goes beyond the data; that is, it requires a generalization.
 (b) Since 12,305 exceeds the other four figures, the conclusion is purely descriptive.
 (c) Since 12,305 and 11,733 exceed 11,000 while the other three figures do not, the conclusion is purely descriptive.
 (d) Since there may have been other reasons for the change in attendance (maybe the fourth game coincided with homecoming), the conclusion entails a generalization.

1.17 (a) Persons walking near a government office may have some relationship to the government and thus have non-typical opinions.
 (b) The use of the word "denied" introduces a bias.

1.19 (a) "Indian art" may be interpreted as art from India or as art of American Indians.
 (b) Many persons will not respond honestly to questions about their personal habits.

1.21 The data are only ordinal data and should not have been added. One way of comparing the overall performance of the two golfers is to compare their total prize money.

1.23 (a) Descriptive. The conclusion follows from the data.
 (b) Descriptive. The conclusion follows from the data.
 (c) Generalization. There may have been other reasons which lead to the conclusion.
 (d) Generalization. There may have been other reasons which lead to the conclusion.

CHAPTER 2

Summarizing Data: Frequency Distributions

2.1
```
15 | 8
16 | 13667
17 | 0122478
18 | 2                NOTE: 15|8 means 158
19 | 1
20 | 1
```

2.3
```
3 | 29
4 | 2456888
5 | 0126
6 | 12               NOTE: 3|2 means $320
```

In this display, the leaves are *tens* digits. The data values were not rounded.

2.5
```
16 | 4
16 |
17 | 02
17 | 68
18 | 0123
18 | 668             NOTE: 16|4 means 164
19 | 334
19 | 8
20 |
20 | 58
21 | 0
21 |
22 |
22 | 5
```

2.7
```
4 | 027
5 | 135889
6 | 0123679          NOTE: 4|0 means 40
7 | 123
8 | 0
```

2.9 Many schemes are possible, but the simplest is probably this: 160–169, 170–179, 180–189, 190–199, 200–209, 210–219, 220–229, 230–239, 240–249, 250–259, and 260–269. This assumes that the weights are given to the nearest pound.

2.11 One possible scheme: 148.0–149.9, 150.0–151.9, 152.0–153.9, 154.0–155.9, 156.0–157.9, 158.0–159.9, and 160.0–161.9. The class width is 2°, and the intervals begin at 148°, 150°, 152°, and so on.

2.13 (a) yes; (b) no; (c) yes; (d) no; (e) no.

2.15 (a) 89; (b) 36;
(c) through (f) cannot be determined exactly from the given information.

2.17 (a) The class limits are the values as given; namely $0.00–49.99, $50.00–99.99, $100.00–149.99, $150.00–199.99, $200.00–249.99, $250.00–299.99, and $300.00 and over.
(b) The class boundaries are –$0.005, $49.995, $99.995, $149.995, $199.995, $249.995, and $299.995. Note that the final class has no upper boundary. We use –$0.005 as the lower limit of the first class in order to maintain an interval of $50 for this class.
(c) The class marks are $24.995, $74.995, $124.995, $174.995, $224.995, and $274.995. The final class does not have a class mark.
(d) The class interval is $50, except for the final class, which has no upper bound.

2.19 The reported values must be whole numbers. Observe that the class interval must be 13 = 19 – 6 = 32 – 19 = 45 – 32.
(a) Since the boundaries must be halfway between the class marks, these are –0.5, 12.5, 25.5, 38.5, and 51.5.
(b) The class limits are 0–12, 13–25, 26–38, and 39–51.

2.21 The second and third classes overlap, in that values from $35.00 to $35.99 could be placed in either of these classes. Also, there is no provision for values from $49.91 to $49.99.

2.23

Grades	Percentage
30–39	7.5%
40–49	5.0%
50–59	5.0%
60–69	25.0%
70–79	30.0%
80–89	17.5%
90–99	10.0%
Total	100.0%

2.25

Weights	Frequency
80–89	2
90–99	3
100–109	9
110–119	13
120–129	13
130–139	7
140–149	3
Total	50

2.27

Weights	Frequency
80 or more	50
90 or more	48
100 or more	45
110 or more	36
120 or more	23
130 or more	10
140 or more	3
150 or more	0

2.29 (a)

Number of Customers	Percentage
40–44	3.33%
45–49	7.50%
50–54	15.00%
55–59	27.50%
60–64	31.67%
65–69	10.83%
70–74	2.50%
75–79	1.67%
Total	100.00%

2.29 (b)

Number of Customers	Percentage
Less than 40	0.00%
Less than 45	3.33%
Less than 50	10.83%
Less than 55	25.83%
Less than 60	53.33%
Less than 65	85.00%
Less than 70	95.83%
Less than 75	98.33%
Less than 80	100.00%

2.31 (a)

Miles per gallon	Frequency
More than 22.4	40
More than 22.9	38
More than 23.4	35
More than 23.9	28
More than 24.4	20
More than 24.9	6
More than 25.4	1
More than 25.9	0

Since the data values were rounded to the nearest tenth, it would be reasonable also to name the classes as "More than 22.45," "More than 22.95," and so on...

2.31 (b)

Miles per gallon	Percentage
Less than 22.4	100.0%
Less than 22.9	95.0%
Less than 23.4	87.5%
Less than 23.9	70.0%
Less than 24.4	50.0%
Less than 24.9	15.0%
Less than 25.4	2.5%
Less than 25.9	0.0%

2.33

Number of mistakes	Frequency
0	17
1	21
2	8
3	7
4	5
5	2
Total	60

2.35

Means of Transportation	Frequency
Bus	8
Car	18
Plane	10
Train	4
Total	40

2.37 For instance, defective merchandise could include either merchandise of incorrect color or size or be damaged during delivery.

2.39

Amount in Dollars	Frequency
Less than 20.00	22
Less than 40.00	69
Less than 60.00	135
Less than 80.00	170
Less than 100.00	191
Less than 120.00	200

2.41

Weight (pounds)	Frequency
Less than 90	0
Less than 100	6
Less than 110	31
Less than 120	77
Less than 130	114
Less than 140	136
Less than 150	143
Less than 160	146
Less than 170	149
Less than 180	149
Less than 190	150

2.43

Weight (Pounds)	Frequency
Less than 160	0
Less than 170	2
Less than 180	8
Less than 190	20
Less than 200	33
Less than 210	51
Less than 220	68
Less than 230	75
Less than 240	83
Less than 250	87
Less than 260	88
Less than 270	90

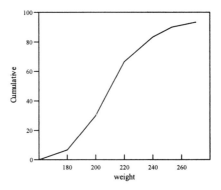

2.45 It gives a misleading impression because we tend to compare areas rather than the heights of the rectangles. We can make the areas represent the class frequencies by dividing the height of the fourth rectangle by 2.

2.47 The central angles of the six sectors are 209°, 52°, 35°, 10°, 46°, and 8°, rounded to the nearest degree.

2.49

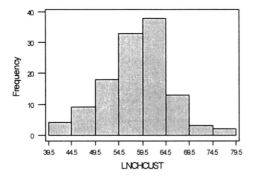

2.51 15–21, 22–28, 29–35, 36–42, 43–49, and 50–56.

2.53 The class frequencies are 7, 10, 20, 22, 10, 2, and 1.

2.55 The cumulative "or less" frequencies are 7, 17, 37, 59, 69, 71, and 72.

2.57 (a) No; (b) yes; (c) yes; (d) no.

2.59 The "or more" cumulative percentages are 100, 73.33, 31.67, 10, 3.33, and 0.

2.61 A good choice would be 18,000–20,499, 20,500–22,999, 23,000–25,499, 25,500–27,999, 28,000–30,499, and 30,500–32,999.

2.63 6.0–7.9, 8.0–9.9, 10.0–11.9, 12.0–13.9, and 14.0–15.9.

2.65 (a) 1,253, 1,250, 1,254, 1,258, 1,257, 1,256, and 1,255;
(b) 3,467, 3,405, 3,419, and 3,448.
(c) 11, 10, 18, 16, 17, 17, 22, 24, 20, 21, 21, 23, 29, 26, 27, 27, 33, 31, 32, 38, 35, and 42.

2.69 Some may not have finished high school.

CHAPTER 3

Summarizing Data: Measures of Location

3.1 (a) This would be regarded as a population if we are interested only in this single company.
(b) This would be regarded as a sample if we are interested generally in the expenses of computer support groups of many companies.

3.3 (a) These would be regarded as a population if the set of 23 department stores constitutes a single chain of stores run under one central management.
(b) These would be regarded as a sample if these were 23 stores out of a larger chain or if the 23 stores were completely unrelated.

3.5 The mean is $\frac{1,254}{30} = 41.8$.

3.7 (a) The mean is $\frac{143}{12} = 11.9$.
(b) Nine of the drivers were fined $60, and three were fined $88. The total fine amount was $9 \cdot \$60 + 3 \cdot \$88 = \$804$, so that the average fine was $\frac{\$804}{12} = \$67$.

3.9 The mean age is $\frac{581}{10} = 58.1$.

3.11 The 18 vehicles have a total weight of 83,340 pounds. The bridge is quite safe with this weight.

3.13 The total value of all slips is $72. The total of the nine slips in hand is $65.10, so that the missing slip must be $6.90.

3.15 (a) The mean amount is $\frac{24.06}{6} = 4.01$ ounces.
(b) The mean would have been calculated as $\frac{24.51}{6} = 4.085$ ounces. This is an error of $4.085 - 4.01 = 0.075$ ounce.

3.17 (a) The fraction of dogs with weight exceeding 40 pounds can be at most $\frac{35}{40} = \frac{7}{8}$.
(b) The fraction of trees with a diameter of 24 cm or more can be at most $\frac{16}{24} = \frac{2}{3}$.

3.19 (a) 1.6.
(b) 4.
(c) The two growth rates are $\frac{3}{2}$ and $\frac{8}{3}$, and the geometric mean of these rates is 2. This would lead to a prediction of $2 \cdot 48 = 96$ cases on the fourth day and $2 \cdot 96 = 192$ cases on the fifth day.

3.21 (a) The harmonic mean of 60 and 30 is indeed 40.

(b) The numbers of shares purchased on the two occasions are $\frac{\$18,000}{\$45} = 400$ and $\frac{\$18,000}{\$36} = 500$. The investor has purchased 900 shares for $36,000, so that the average price per share is $\frac{\$36,000}{900} = \$40$. The harmonic mean of $45 and $36 is indeed $40.

(c) The average cost per pound is the harmonic mean of 60, 72, and 90. This is $\frac{3}{\frac{1}{60} + \frac{1}{72} + \frac{1}{90}} = 72$.

3.23 The average salary is
$$\frac{(720 \times \$38,300) + (660 \times \$44,500) + (520 \times \$41,000)}{720 + 660 + 520} = \frac{\$78,266,000}{1,900} = \$41,192.63.$$

3.25 The sample mean is $\frac{80 \times 0.9 + 70 \times 0.7 + 65 \times 1.0}{80 + 70 + 65} = \frac{186}{215} \approx 0.865$ hour.

3.27 (a) Since $\frac{25+1}{2} = 13$, the median is the 13th value.

(b) Since $\frac{32+1}{2} = 16.5$, the median is the mean of the 16th and 17th values.

3.29 Arranged according to size, the data are 37, 38, 40, 40, 46, 48, 50, 52, 55, 56, 56, 58, 60, 61, and 63. Since $\frac{15+1}{2} = 8$, the median is the 8th value, namely, 52.

3.31 Arranged according to size, the data are 12, 18, 26, 28, 31, 33, 40, 44, 45, 49, 61, 63, 75, 80, 80, 89, 96, 103, 125, and 125. Since $\frac{20+1}{2} = 10.5$, the median is the mean of the 10th and 11th values, namely $\frac{49+61}{2} = 55$.

3.33 (a) $\bar{x} = \frac{0+0+1+2+0+3+1+0+0+0+0+1+0+0+\cdots}{20} = 0.8$.

(b) Since $\frac{20+1}{2} = 10.5$, the median is the mean of the 10th and 11th values, namely 0.

(c) 9 values are above the mean.
11 values are below the mean.

(d) 9 values are above the median.
No value is below the median.

3.35 (a) Here are the data arranged according to size:

113	118	121	123	126
128	130	135	137	138
139	140	140	142	142
142	142	143	146	155
157	157	158	159	164

It is easily seen that the median (the 13th value) is 140 minutes.

(b) The steam-and-leaf display is this:

```
11 | 38
12 | 1368       NOTE: 12|1 means 121
13 | 05789
14 | 00222236
15 | 57789
16 | 4
```

Counting to the 13th position will reveal the median to be 140 minutes.

3.37 $\bar{x} = \dfrac{107+90+80+\ldots+95+91}{15} = \dfrac{1{,}657}{15} \approx 110.5\%$. Arranged according to size, the data are 74, 78, 80, 86, 90, 91, 92, 92, 95, 102, 102, 106, 107, 109, and 353. Since $\dfrac{15+1}{2} = 8$, the median is the 8th value, namely, 92%. The mean is influenced strongly by the single person at 353% of quota, and the median is a better indicator of "average" performance.

3.41 Since the three midranges are 29.8, 30.0, and 30.3, the manufacturers of car C can use the midrange to substantiate the claim that their car performed best.

3.43 It is easiest to begin by making a stem-and-leaf display:

```
4 | 6
5 | 56888
6 | 11567889
7 | 000245
8 | 0
```

(a) The median is the 11th value, namely 67.

(b) Q_1 is the average of the 5th and 6th values, which is $\dfrac{58+58}{2} = 58$.

Q_3 is the average of the 17th and 16th values, which is $\dfrac{70+70}{2} = 70$.

3.45 Here are the positions:

	n	Median	Lower Quartile	Upper Quartile
(a)	40	20.5	10.5	30.5
(b)	41	21.0	10.5	31.5
(c)	42	21.5	11.0	32.0
(d)	43	22.0	11.0	33.0

Half-integers refer to the means of the values in the surrounding positions; for example, 21.5 refers to the average of the 21st and 22nd values.

3.47 Here is a stem-and-leaf display:

```
 1 | 28
 2 | 68
 3 | 13
 4 | 0459
 5 |
 6 | 13        Note 1|2 means 12
 7 | 5
 8 | 009
 9 | 6
10 | 3
11 |
12 | 55
```

Q_1 is the average of the 5th and 6th values, which is $\frac{31+33}{2} = 32$.

Q_3 is the average of the 15th and 16th values, which is $\frac{80+89}{2} = 84.5$.

3.49 To find the 30th percentile position in a list of 21, we find $0.3(21) = 6.3$ and use the mean of the 6th and 7th values, which is $\frac{58+61}{2} = 59.5$.

3.53 The mode is 7, which occurs three times.

3.55
(a) The mode is 146, which occurs four times.
(b) The mode is 149, which occurs four times.
(c) Each value in the list occurs just once, so that there is no mode.

3.57
(a) The mean is 581.2, the median is 591.5, and the mode is 514.
(b) The mean is 580, the median is 590, and the mode is 480.
(c) The mean is 585, the median is 600, and the mode is 600.
(d) The mean and median are altered only modestly by rounding. The mode can change radically when data are rounded. The mode must be used with extreme caution with measured data.

3.59 $\bar{x} = \frac{17 \cdot 44 + 32 \cdot 70 + 47 \cdot 92 + 62 \cdot 147 + 77 \cdot 115 + 92 \cdot 32}{500} = \frac{28,225}{500} = 56.45$.

3.61
(a) $\tilde{x} = \$0.00 + \frac{14}{66} \times \$50.00 \approx \$10.61$.

(b) $P_{20} = -\$50 + \frac{9}{31} \times \$50.00 \approx -\$35.48$.

$P_{80} = \$0.00 + \frac{50}{66} \times \$50.00 \approx \$37.88$.

3.63 $\bar{x} = \frac{2 \cdot 18 + 7 \cdot 15 + 12 \cdot 9 + 17 \cdot 7 + 22 \cdot 1}{50} = \frac{390}{50} = 7.8$.

3.65
(a) $D_3 = -0.5 + \frac{15}{18} \cdot 5 \approx 3.7$ and $D_9 = 14.5 - \frac{7}{9} \cdot 5 \approx 10.6$;

(b) $P_5 = -0.5 + \frac{2.5}{18} \cdot 5 \approx 0.2$ and $P_{98} = 19.5$.

3.67 (a) $\tilde{x} = 34.50 + \dfrac{106.50}{178} \cdot 5 \approx 37.49$

(b) $P_{20} = 29.50 + \dfrac{5.80}{115} \cdot 5 \approx 29.75$

$P_{40} = 34.50 + \dfrac{34.60}{178} \cdot 5 \approx 35.47$

$P_{60} = 39.50 + \dfrac{0.40}{107} \cdot 5 \approx 39.52$

$P_{80} = 44.50 + \dfrac{37.20}{88} \cdot 5 \approx 46.61$

3.69 Using the class marks 84.5, 94.5, and so on, we find the mean of the distribution to be 117.5. The original data values have mean 117.88, so that the grouping error is only $117.5 - 117.88 = -0.38$. By the way, the grouping error is $\dfrac{-0.38}{117.88} \cdot 100\% \approx -0.32\%$ as a percentage of the mean.

3.71 For Exercise 3.69, the class interval is 10, and the grouping error is –0.38. For Exercise 3.70, the class interval is 5, while the grouping error is –0.0075. In each case, the grouping error is much less than the class interval.

3.73 (a) $x_1 + x_2 + x_3 + x_4 + x_5 + 5$;

(b) $3(y_1 + y_2 + y_3 + y_4)$;

(c) $3(x_1 + x_2 + x_3 + x_4)$.

3.75 (a) 25; (b) 14; (c) 70; (d) 430.

3.77 (a) For $j=1$, the sum is 7;
For $j=2$, the sum is 4;
For $j=3$, the sum is –2;
For $j=4$, the sum is 10.

(b) For $i=1$, the sum is 4;
For $i=2$, the sum is 8;
For $i=3$, the sum is 7.

3.81 (a) $\sum_{i=1}^{n} a_i^2 = 5^2 + 3^2 + 2^2 + 1^2 = 39 < \left(\sum_{i=1}^{n} a_i\right)^2 = (5+3+2+1)^2 = 121$

(b) $\sum_{i=1}^{n} a_i^2 = 1^2 + 2^2 + 0^2 + 1^2 + 6^2 = 42 < \left(\sum_{i=1}^{n} a_i\right)^2 = (1+2+0+1+6)^2 = 100$

(c) Take $n = 2$, $a_1 = 1$, $a_2 = -1$; $\sum_{i=1}^{n} a_i^2 = 1+1 = 2 > \left(\sum_{i=1}^{n} a_i\right)^2 = (1-1)^2 = 0$

3.83 (a) $\sum_{i=1}^{n} |x_i||y_i| = 2 \cdot 100 + 3 \cdot 200 + 5 \cdot 300 = 2,300$

$< \sum_{i=1}^{n} |x_i| \sum_{i=1}^{n} |y_i| = (2+3+5)(100+200+300) = 6,000$

(b) $\sum_{i=1}^{n} |x_i||y_i| = 0 \cdot 0 + 2 \cdot 20 + 0 \cdot 0 + 4 \cdot 50 = 240$

$< \sum_{i=1}^{n} |x_i| \sum_{i=1}^{n} |y_i| = (0+2+0+4)(0+20+0+50) = 420$

3.85 The median is the average of the 8th and 9th values, which is $\frac{1.57+1.58}{2} = 1.575$.

Q_1 is the average of the 4th and 5th values, which is $\frac{1.36+1.39}{2} = 1.375$.

Q_3 is the average of the 12th and 13th values, which is $\frac{1.72+1.76}{2} = 1.74$.

3.87 $\bar{x} = \frac{6 \cdot 5 + 9 \cdot 9 + 12 \cdot 12 + 15 \cdot 8 + 18 \cdot 13 + 21 \cdot 3}{60} = \frac{822}{60} \approx 13.7$.

3.89 The 60th percentile for this set of 60 values is $\frac{10}{18}$ of the way through the class of 14–16. Its value can be found as $13.5 + \frac{10}{18} \cdot 3 \approx 15.17$.

3.91 (a) The median is in position 16.
(b) The median is the mean of the values in positions 40 and 41.

3.93 The median is the average of the 5th and 6th values, which is $\frac{27.3+28.2}{2} = 27.75$.

Q_1 is the third value, namely 25.8.
Q_3 is the 8th value, namely 30.5.

3.95 The total number of store visits by the twelve persons is $12 \cdot 5.75 = 69$. It is therefore not possible that seven or more of these persons shopped in ten or more stores.

3.97 The median is in the 12th position. The lower quartile is in the 6th position. The upper quartile is in the 18th position.

3.99 Here is the stem-and-leaf display using unrounded whole numbers.

```
1 | 7899
2 | 1458                NOTE: 2|1 means 21
3 | 12333477899
4 | 11222333455789
5 | 1334679
6 | 06
```

The median is the mean of the 21st and 22nd values; it is equal to 41.5. If you go back to the original list to locate the median, you could give its value as 41.95.

3.101 (a) $Q_1 = 15.5 + \frac{6 \cdot 5}{8} \cdot 5 \approx 19.6$; (b) $Q_3 = 30.5 - \frac{2 \cdot 5}{13} \cdot 5 \approx 29.5$;

(c) $P_{40} = 20.5 + \frac{5.4}{11} \cdot 5 = 23.0$; (d) $P_{80} = 25.5 - \frac{12.8}{13} \cdot 5 = 30.4$.

3.103 It is possible that the mean salary is higher at company B. Here is a very simple situation in which this can happen:

Company	Male Employees Number	Mean Salary	Female Employees Number	Mean Salary
A	1	$30,000	2	$24,000
B	2	$28,000	1	$23,000

The mean salary at company A is $26,000. The mean salary at company B is $26,333.

CHAPTER 4

Summarizing Data: Measures of Variation

4.1 The range is 14 − 6 = 8 seconds.

4.3 The range is $16\frac{5}{8} - 15\frac{1}{2} = 1\frac{1}{8}$ for stock A and $22\frac{1}{4} - 21\frac{7}{8} = \frac{3}{8}$ for stock B.

4.5 The quartiles are the means of the 6th and 7th values from the ends of the sorted list. Numerically, these are 15 and 17.5. The interquartile range is then 2.5. It should not be regarded as surprising that the interquartile range is less than half the range.

4.7 The lower quartile is the median of all values below the median position. In this problem, it is the median of the lowest 10 values and is found as the mean of the fifth and sixth values; numerically it is 58. The upper quartile is found in a similar manner from the high end, and its value is 70. The interquartile range is 70 − 58 = 12 feet.

4.9 The lower quartile is the 13th value from the low end, namely 53. The upper quartile is the 13th value from the high end, namely 78. The semi-interquartile range is $\frac{1}{2}(78 - 53) = 12.5$.

4.11 (a) Begin by finding the total $\Sigma x = 64$ and the average $\bar{x} = 16$. Then
$$s^2 = \frac{(16.2-16)^2 + (15.9-16)^2 + (15.8-16)^2 + (16.1-16)^2}{4-1} \approx 0.0333.$$ Then $s = \sqrt{0.0333} \approx 0.18$.

(b) Using $\Sigma x = 64$ and $\Sigma x^2 = 1{,}024.1$, find $s^2 = \dfrac{1{,}024.1 - \frac{(64)^2}{4}}{4-1} \approx 0.0333$. Then $s = 0.18$, as above.

4.13 (a) Using $\Sigma x = 145$ and $\Sigma x^2 = 5{,}299$, find that $s^2 = \dfrac{5{,}299 - \frac{(145)^2}{4}}{4-1} = 14.25$. Then $s = \sqrt{14.25} \approx 3.77$.

(b) After subtracting 30, the list is this: 7, 2, 5, and 11. For this revised list $\Sigma x = 25$ and $\Sigma x^2 = 199$. Then $s^2 = \dfrac{199 - \frac{(25)^2}{4}}{4-1} = 14.25$. This is, of course, identical with the result of part (a). Sometimes the subtraction of a number can reduce the number of figures in the data, and this can greatly reduce the difficulty of the arithmetic.

4.15 Suppose that the list contains $2m$ items. Suppose further that m of these are equal to $a - \frac{b}{2}$. The average will be a and then $s = \sqrt{\frac{m((a-\frac{b}{2})-a)^2 + m((a+\frac{b}{2})-a)^2}{2m-1}} = \sqrt{\frac{m\frac{b^2}{2}}{2m-1}}$. It follows that $s = b\sqrt{\frac{m}{2(2m-1)}}$. In this expression, b is the difference between the two numbers on the list.

(a) Here $s = 4\sqrt{\frac{2}{2(3)}} \approx 2.31$.

(b) Here $s = 100\sqrt{\frac{3}{2(5)}} \approx 54.77$.

(c) Here $s = 20\sqrt{\frac{4}{2(7)}} \approx 10.69$.

The results above were rounded to two decimals. These may be obtained, of course, with the conventional formulas.

4.17 For these data the range is $6 - 2 = 4$ and the standard deviation is 2.06, rounded to two decimals. The claim is certainly reasonable for this set of values.

4.19 (b) After subtracting 10 from each value, we get $\Sigma x = 159$ and $\Sigma x^2 = 1,097$, so that

$$s = \sqrt{\frac{1,097 - \frac{(159)^2}{25}}{24}} = 1.89.$$

4.21 (a) The proportion is at least $1 - \frac{1}{6^2} = \frac{35}{36}$.

(b) The proportion is at least $1 - \frac{1}{12^2} = \frac{143}{144}$.

4.23 Since $k = 0.5 < 1$, Chebyshev's theorem won't work in this case.

4.25 The percentage is at least

(a) 96%; (b) 98.4%; (c) 99%; (d) 99.75%

4.27 (a) Since $1 - \frac{1}{k^2} = \frac{35}{36}, \frac{1}{k^2} = \frac{1}{36}$ and $k = 6$, and the thiamine content of $\frac{35}{36}$ of the slices must be between $0.260 - 6(0.005) = 0.230$ and $0.260 + 6(0.005) = 0.290$ milligram.

(b) Since $1 - \frac{1}{k^2} = \frac{80}{81}, \frac{1}{k^2} = \frac{1}{81}$ and $k = 9$, and the thiamine content of $\frac{80}{81}$ of the slices must be between $0.260 - 9(0.005) = 0.215$ and $0.260 + 9(0.005) = 0.305$ milligram.

Modern Elementary Statistics – 9th Edition 17

4.29 (a) MTB > Mean C1

Column Mean

Mean of LNCHCUST = 58.400

MTB > Stan C1

Column Standard Deviation

Standard deviation of LNCHCUST = 6.9299

(b) 88/120 = 73.33% of the data values lie within one standard deviation of the mean (58.4 ± 1[6.9299]).

112/120 = 93.33% of the data values lie within two standard deviations of the mean (58.4 ± 2[6.9299]).

120/120 = 100.0% of the data values lie within three standard deviations of the mean (58.4 ± 3[6.9299]).

According to the rule on pages 87 - 88, about 68% of the data values should lie within one standard deviation of the mean, about 95% within two standard deviations of the mean, and about 99.7% within three standard deviations of the mean.

4.31 The z-score for the first man is $\frac{178-145}{15} = 2.2$, and the z-score for the second man is $\frac{204-165}{20} = 1.95$. By this standard, the first man is more seriously overweight. You might find it interesting to note that the first man is over his target weight by $\frac{33}{145} \approx 23\%$ while the second man is over his target by $\frac{39}{165} \approx 24\%$.

4.33 The coefficient of variation for the first patient is $\frac{14.2}{188} \approx 0.0755 = 7.55\%$. The coefficient of variation for the second patient is $\frac{8.6}{136} \approx 0.0632 = 6.32\%$. This suggests that the first patient's blood pressure is more variable. The conditions of these patients should not be judged on the coefficient of variation alone. The first patient is seriously ill.

4.35 Since $Q_1 = 58$ and $Q_3 = 70$, we have $\frac{70-58}{70+58} \cdot 100\% \approx 9.4\%$.

4.37 (a) $\bar{x} = \frac{4.5 \cdot 2 + 14.5 \cdot 15 + 24.5 \cdot 17 + 34.5 \cdot 13 + 44.5 \cdot 3}{50} = \frac{1,225}{50} = 24.5$

and $\bar{x} = 19.5 + \frac{8}{17} \cdot 10 \approx 24.21$.

(b) Since $\Sigma x \cdot f = 1,225$ and $\Sigma x^2 \cdot f = 34,812.5$, the standard deviation is

$$s = \sqrt{\frac{34,812.5 - \frac{1,225^2}{50}}{50-1}} \approx 9.90.$$

4.39 $SK = \dfrac{3(150.68 - 149.30)}{13.79} = 0.30$.

4.41 $SK = \dfrac{3(56 \cdot 45 - 59.0)}{20.62} \approx -0.37$. The distribution is fairly symmetrical.

4.45 The smallest value is 46, the lower hinge is 58, the median is 67, the upper hinge is 70, and the largest value is 80. We get

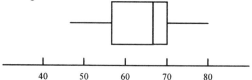

Distribution is fairly symmetrical.

4.47 The frequencies corresponding to 0, 1, 2 and 3 are 28, 17, 4 and 1; the data are highly skewed with the tail on the right-hand side.

4.49

numbers of loaves of bread unsold

The data are highly skewed with the tail on the right-hand side.

4.51 The smallest value is 0, the lower hinge is 0, the median is 1, the upper hinge is 5, and the largest value is 5, and we get

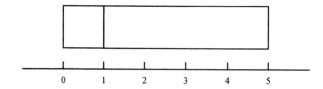

There are no "whiskers" on either side; the data are U-shaped.

4.53 $k = 1.5$, $1 - \dfrac{1}{k^2} = 0.56$.

4.55 The values show a very slight positive skewness.

4.57 The value of σ^2 is 2, and this is precisely $\dfrac{5^2 - 1}{12}$, so that the claim is supported for this case. (This is not a proof that the claim is true.)

4.59 The standard deviation is $s = 28 \cdot 22\% = 6.16$. The standard score for a watermelon that weighs 35 pounds therefore is $\frac{35-28}{6.16} = 1.14$.

4.61 The class marks are 8, 13, 18, 23, and 28

(a) $\bar{x} = \frac{12(8)+73(13)+52(18)+39(23)+24(28)}{200} = \frac{3,550}{200} = 17.75$

(b) From part (a), $\Sigma xf = 3,550$, and $(\Sigma xf)^2 = 3,550^2$;
$x^2 f = 12(8)^2 + 73(13)^2 + 52(18)^2 + 39(23)^2 + 24(28)^2 = 69,400$
$S_{xx} = 69,400 - \frac{3,550^2}{200} = 6,387.7$; and then $s = \sqrt{\frac{6,387.5}{199}} = 5.66$.

4.63 We are given two facts about μ and σ:
$\frac{1,020-\mu}{\sigma} = 2$ and $\frac{\sigma}{\mu} = 0.14$. The facts can be solved (in a variety of ways) to get $\mu = 796.88$ and $\sigma = 111.56$.

4.65 (a) The median is 22.
(b) The quartiles are 12 and 29.

4.67 (a) This interval corresponds to a spread of 1.5 standard deviations about the mean, and at least 55.6% of the scores must be inside the interval.
(b) The interval corresponds to a spread of 2 standard deviations about the mean, and at least 75% of the scores must be inside the interval.
(c) The interval corresponds to a spread of 2.5 standard deviations about the mean, and at least 84% of the scores must be inside the interval.

4.69 The variance is 103.74.

4.71 By Chebyshev's theorem, we can get $k = 2$, $\bar{x} = 265$, $s = 10$. Therefore, the coefficient of variation is $V = \frac{10}{265} \cdot 100\% \approx 3.77\%$.

CHAPTER 5

Possibilities and Probabilities

5.1 In the diagram, an N denotes a win by the National League team, and A denotes a win by the American League team

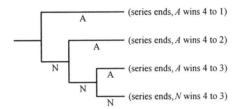

- (series ends, A wins 4 to 1)
- (series ends, A wins 4 to 2)
- (series ends, A wins 4 to 3)
- (series ends, N wins 4 to 3)

5.3 (a)

```
   ┌── 1 ── 2 ── 2 ──
───┤       ┌── 1 ── 2 ──
   └── 2 ──┤
           └── 2 ── 1 ──
```

There are exactly three ways in which the student can study for a total of five hours over the three nights.

(b)

```
   ┌── 1 ── 2 ── 2 ──
───┤       ┌── 1 ── 2 ──
   └── 2 ──┤       ┌── 1 ──
           └── 2 ──┤
                   └── 2 ──
```

There are four ways in which the student can study for a total of at least five hours over the three nights.

5.5

Day 1   Day 2   Day 3

```
        ┌── 0 ──┬── 2 ── 3 ──
        │       └── 3 ── 2 ──
        │       ┌── 1 ── 3 ──
        ├── 1 ──┼── 2 ── 2 ──
        │       └── 3 ── 1 ──
────────┤       ┌── 0 ── 3 ──
        │       ├── 1 ── 2 ──
        ├── 2 ──┼── 2 ── 1 ──
        │       └── 3 ── 0 ──
        │       ┌── 0 ── 2 ──
        └── 3 ──┼── 1 ── 1 ──
                └── 2 ── 0 ──
```

There are 12 ways in which the store can sell five cream cakes on three consecutive days.

5.7   There are 30 different ways.

5.9   There are 8 different ways.

5.11   There are $6 \cdot 4 \cdot 2 \cdot 2 = 96$ ways.

5.13   240.

5.15   There are $2^{10} = 1,024$ possible pizzas.

5.17   (a) False;   (b) False;   (c) True;   (d) True.

5.19   If he starts at an Italian restaurant, he has $8 \cdot 7 \cdot 6 \cdot 5 = 1,680$ choices for his dining on nights 1, 3, 5 and 7. He will also have $9 \cdot 8 \cdot 7 = 504$ choices for dining in a Chinese restaurant on nights 2, 4, and 6. There are altogether $1,680 \cdot 504 = 846,720$ ways to plan his evening meal, starting at an Italian restaurant. If he starts at a Chinese restaurant, the number of choices is $(9 \cdot 8 \cdot 7 \cdot 6) \cdot (8 \cdot 7 \cdot 6) = 3,024 \cdot 336 = 1,016,064$. The total number of choices is $846,720 + 1,016,064 = 1,862,784$.

5.21 $10 \cdot 9 \cdot 18 \cdot 17 \cdot 16 \cdot 15 = 6,609,600$.

5.23 $8! = 40,320$.

5.25   (a)   There are $4! = 24$ orders in which to place the four couples. For each couple there are still two choices for positions (husband on left or husband on right), so that the total number of choices is $24 \cdot 2^4 = 384$.

       (b)   There are $4! = 24$ ways to arrange the men and also 24 ways to arrange the women. In addition, we have to decide whether the group of men sits on the left or on the right. The total number of arrangements of this type is $24 \cdot 24 \cdot 2 = 1,152$.

       (c)   Coalesce the men into a single block. Now there are $5! = 120$ ways to arrange the four women and the block of men. Internally, this block of men can be arranged in $4! = 24$ ways, so that the total number of possible arrangements of this type is $120 \cdot 24 = 2,880$.

       (d)   There are $4! \cdot 4! = 576$ arrangements if the men sit in the odd-numbered seats. To this we add the 576 arrangements in which the women sit in the odd-numbered seats. Altogether there are 1,152 arrangements of this type.

       (e)   It is important to realize that this problem does not allow a woman to sit next to another woman.

5.27   (a) $\dfrac{6!}{2! \cdot 2!} = 180$;   (b) $\dfrac{6!}{3! \cdot 3!} = 60$;   (c) $\dfrac{6!}{3! \cdot 3!} = 20$;

       (d)   Suppose that among the n objects, there are k different types, with r_1 of the first type, r_2 of the second type, r_3 of the third type, and so on, up through r_k of the kth type. Observe that $r_1 + r_2 + \cdots + r_k = n$ and that some of the r-values might be equal to 1. The number of arrangements is then $\dfrac{n!}{r_1! \cdot r_2! \cdot r_3! \cdot \cdots \cdot r_k!}$.

       (e) $\dfrac{11!}{1! \cdot 4! \cdot 4! \cdot 2!} = 34,650$. The 1! (for the letter M) may be omitted in the calculation.

5.29 (a) The position of the first object is irrelevant. Once this is in place, there are $(n-1)!$ ways that the other objects can be arranged relative to it.
(b) $5! = 120$.
(c) $7! = 5,040$.
(d) Arrange the men and women separately. There are $3! = 6$ possible arrangements for each. Once these arrangements have been selected, there are 4 ways to put them together; consider which woman should be placed to the right of Mr. A. Altogether there are $6 \cdot 6 \cdot 4 = 144$ arrangements.

5.31 $\binom{15}{4} = 1,365$ **5.33** $\binom{10}{4} = 210$ **5.35** $\binom{15}{5} = 3,003$

5.37 (a) $\binom{8}{3} \cdot \binom{2}{0} = 56 \cdot 1 = 56;$ (b) $\binom{8}{2} \cdot \binom{2}{1} = 28 \cdot 2 = 56;$
(c) $\binom{8}{1} \cdot \binom{2}{2} = 8 \cdot 1 = 8.$

5.39 $\binom{8}{2} \cdot \binom{6}{2} \cdot \binom{10}{2} = 28 \cdot 15 \cdot 45 = 18,900.$

5.41 (a) $(a+b)^1 = a + b = \binom{1}{0}a + \binom{1}{1}b.$
(b) $(a+b)^2 = a^2 + 2ab + b^2 = \binom{2}{0}a^2 + \binom{2}{1}ab + \binom{2}{2}b^2.$ Note that $\binom{2}{0} = 1$, $\binom{2}{1} = 2$, and $\binom{2}{2} = 1.$
(c) $(a+b)^3 = a^3 + 3a^2b + 3ab^2 + b^3 = \binom{3}{0}a^3 + \binom{3}{1}a^2b + \binom{3}{2}ab^2 + \binom{3}{3}b^3.$ Note that $\binom{3}{0} = 1$, $\binom{3}{1} = 3$, $\binom{3}{2} = 3$, and $\binom{3}{3} = 1.$
(d) $(a+b)^4 = a^4 + 4a^3b + 6a^2b^2 + 4ab^3 + b^4 = \binom{4}{0}a^4 + \binom{4}{1}a^3b + \binom{4}{2}a^2b^2 + \binom{4}{3}ab^3 + \binom{4}{4}b^4.$
Note that $\binom{4}{0} = 1$, $\binom{4}{1} = 4$, $\binom{4}{2} = 6$, $\binom{4}{3} = 4$, and $\binom{4}{4} = 1.$
(e) The results agree with Table X.

5.43 (a) From Table X on page 607 we get $\binom{12}{8} = 495$, $\binom{11}{8} = 165$, $\binom{14}{4} = 1,001$, $\binom{13}{4} = 715$, $\binom{10}{4} = 210$, and $\binom{9}{3} = 84$. Substituting these values into the equations, we get (a) $495 = 3(165)$, and $495 = 495$; (b) $10(1,001) = 14(715)$, and $1,010 = 1,010$; and (c) $\frac{210}{84} = \frac{5}{2}$, and $2.5 = 2.5$.

5.45 (a) Row 7 is 1 7 21 35 35 21 7 1.

(b) $\binom{7}{3} = 35$.

(c) The right side is
$$\frac{(n-1)!}{(r-1)!(n-r)!} + \frac{(n-1)!}{r!(n-1-r)!} = \frac{(n-1)!}{(r-1)!(n-1-r)!}\left(\frac{1}{n-r} + \frac{1}{r}\right)$$
$$= \frac{(n-1)!}{(r-1)!(n-1-r)!}\left(\frac{n}{(n-r)r}\right) = \frac{n!}{r!(n-r)!},$$
which is the desired left side.

5.47 (a) $\binom{7}{3} = 35$; (b) $\binom{6}{3} = 20$;

(c) $\binom{6}{2} = 15$, the number of ways of choosing the remaining two people, once Susan is set aside;

(d) easily verified, $20 + 15 = 35$.

5.49 (a) $\dfrac{\binom{4}{2}}{\binom{52}{2}} = \dfrac{4 \cdot 3}{52 \cdot 51} \cdot \dfrac{2}{2} = \dfrac{3}{663} = \dfrac{1}{221}$; (b) $\dfrac{\binom{12}{2}}{\binom{52}{2}} = \dfrac{12 \cdot 11}{52 \cdot 51} \cdot \dfrac{2}{2} = \dfrac{11}{221}$;

(c) $\dfrac{\binom{26}{2}}{\binom{52}{2}} = \dfrac{26 \cdot 25}{52 \cdot 51} \cdot \dfrac{2}{2} = \dfrac{25}{102}$; (d) $\dfrac{\binom{13}{2}}{\binom{52}{2}} = \dfrac{13 \cdot 12}{52 \cdot 51} \cdot \dfrac{2}{2} = \dfrac{1}{17}$.

5.51 (a) $\dfrac{1}{6}$; (b) $\dfrac{3}{6} = \dfrac{1}{2}$.

5.53 The probabilities of 0, 1, 2, 3 and 4 heads are, respectively, $\dfrac{1}{16}, \dfrac{4}{16}, \dfrac{6}{16}, \dfrac{4}{16}$, and $\dfrac{1}{16}$.

5.55 (a) $\dfrac{25}{25+18+11} = \dfrac{25}{54}$; (b) $\dfrac{18+11}{25+18+11} = \dfrac{29}{54}$;

(c) $\dfrac{11}{54}$; (d) $\dfrac{18}{54} = \dfrac{1}{3}$.

5.57 (a) $\dfrac{1}{n}$.

5.59 (a) $\dfrac{\binom{22}{2}}{\binom{24}{2}} = \dfrac{231}{276} = 0.837$; (b) $\dfrac{\binom{2}{1}\binom{22}{1}}{276} = \dfrac{44}{276} = 0.159$;

(c) $\dfrac{\binom{2}{0}\binom{22}{0}}{276} = \dfrac{1}{276} = 0.004$.

5.61 (a) $\dfrac{\binom{1}{1}\cdot\binom{6}{1}}{\binom{7}{2}} = \dfrac{1\cdot 6}{21} = \dfrac{2}{7}$; (b) $\dfrac{1}{\binom{7}{2}} = \dfrac{1}{21}$.

5.63 The estimate is $\dfrac{226}{300} \approx 0.753$.

5.65 The estimate is $\dfrac{143}{842} \approx 0.170$.

5.67 The estimate is $\dfrac{62}{325} \approx 0.191$.

5.73 (a) The second customer has $9 \cdot 12 = 108$ choices.
(b) The second customer has $10 \cdot 11 = 110$ choices.

5.75 The estimate is $\dfrac{102}{625} \approx 0.1632$.   **5.77** $\dfrac{4}{36} = \dfrac{1}{9}$.

5.79 $4^{10} = 1{,}048{,}576$.

5.81

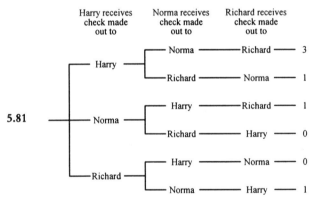

The numbers at the right indicate the number of persons getting the proper check.
(a) 2;   (b) 3;
(c) 0;   (d) 1.

5.83 (a) $8 \cdot 7 \cdot 6 = 336$;   (b) $8 \cdot 7 \cdot 6 = 336$;   (c) $(3!)8 \cdot 7 \cdot 6 = 2{,}016$.

5.85 $\binom{52}{10} = 15{,}820{,}024{,}220$.

5.87 $P(0 \text{ heads}) = \dfrac{1}{32}$, $P(1 \text{ heads}) = \dfrac{5}{32}$, $P(2 \text{ heads}) = \dfrac{10}{32}$, $P(3 \text{ heads}) = \dfrac{10}{32}$, $P(4 \text{ heads}) = \dfrac{5}{32}$,
$P(5 \text{ heads}) = \dfrac{1}{32}$.

5.89 The estimated probability is $\dfrac{684}{800} = \dfrac{81}{100} = 0.81$.

5.91 (a) $\dfrac{286}{560} = 0.511$;   (b) $\dfrac{156}{560} = 0.279$;

(c) $\dfrac{78}{560} = 0.139$;   (d) $\dfrac{1}{560} = 0.002$.

5.93 $24 \cdot 2^5 = 24 \cdot 32 = 768$.

5.95   If you maintain that the identities of the individuals matter, the number of ways is $19 \cdot 18 \cdot 17 \cdot 16 = 93,024$. The number of ways of choosing four entrees out of 19, without regard to the identities of the people eating them, is $\binom{19}{4} = 3,876$.

CHAPTER 6

Some Rules of Probability

6.1 (a) {a, c, d, f, g} is the event that the scholarship is awarded to one of the female applicants.
(b) {a, b, c, e, f, g} is the event that the scholarship is not awarded to Mrs. Daly.
(c) {e} is the event that the scholarship is awarded to Mr. Earl.

6.3 (a) $R' = \{6,7,8,9,10\}$; (b) $R \cap S = \{4,5\}$;
(c) $R' \cap S' = \{8,9,10\}$; (d) $R = \{1,2,3,4,5\}$.

6.5 (a) {8} is the event that a person chooses a color other than red, yellow, blue, green, brown, white, or purple.
(b) This is an empty set.
(c) {1,2,3,4,5} is the event that a person chooses red, yellow, blue, green, or brown.
(d) {3,4,8} is the event that a person chooses blue, green, and another color not red, yellow, brown, white, or purple.

6.7 (a) The first salesman sells one car.
(b) The second salesman sells more cars than the first salesman, or the first salesman sells no cars.
(c) The salesmen sell the same number of cars.
(d) The first salesman sells more cars than the second salesman.

6.9 (a) (0,2), (1,1), (2,0); (b) (2,0), (1,0); (c) (0,0), (2,0), (1,0), (1,1).

6.11 T and U are not mutually exclusive,
T and V are not mutually exclusive,
T and W are mutually exclusive,
U and V are mutually exclusive,
U and W are mutually exclusive,
and V and W are not mutually exclusive.

6.13 (a) The sample space is as follows:

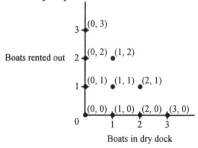

(b) $K = \{(0,2),(0,3),(1,2)\}$,
$L = \{(1,0),(2,0),(2,1),(3,0)\}$, and
$M = \{(0,3),(1,2),(2,1),(3,0)\}$.
(c) K and L are mutually exclusive, K and M are not mutually exclusive, and L and M are not mutually exclusive.

6.15 (a) {A,D}; (b) {C,E}; (c) {B}; (d) {A,C,E}.

6.17 Region 1 represents the event that a driver has both kinds of insurance; Region 2 represents the event that a driver has liability insurance but not collision insurance; Region 3 represents the event that a driver has collision insurance but no liability insurance; Region 4 represents the event that a driver has neither kind of insurance.

6.19 Region 1 represents the event that the flight leaves Denver on time and arrives in Houston on time; Region 2 represents the event that the flight leaves Denver on time but does not arrive in Houston on time; Region 3 represents the event that the flight does not leave Denver on time but arrives in Houston on time; Region 4 represents the event that the flight does not leave Denver on time and does not arrive in Houston on time.

6.21 In the Venn diagram which follows, W denotes women and T denotes players with two-handed backhands:

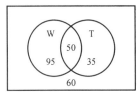

Filling in the numbers associated with the various regions, we find that 60 of the players are men who do not use a two-handed backhand.

6.23 (a) 3,5,6 and 8; (b) 2 and 7; (c) 1 and 3; (d) 7 and 8.

6.25 (a) Region 5; (b) Regions 1 and 2 together;
(c) Regions 3,5, and 6 together; (d) Regions 1,3,4, and 6 together.

6.27

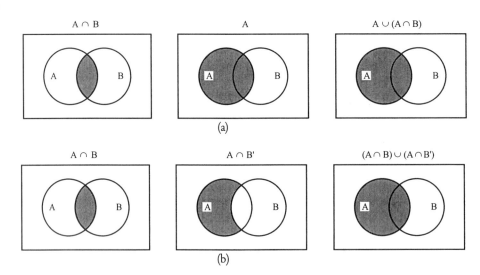

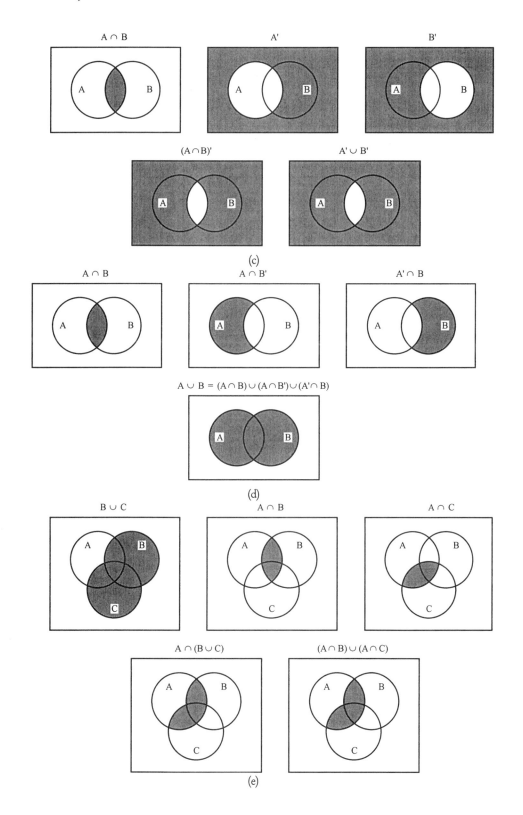

6.29 (a) $P(C' \cap T)$; (b) $P(C' \cap T')$.

6.31 (a) $P(R')$; (b) $P(D' \cap R)$;
(c) $P(V' \cup D')$.

6.33 (a) The third rule; (b) the first rule;
(c) the second rule; (d) the first rule.

6.35 (a) The third rule;
(b) the second probability; cannot exceed the first;
(c) events are not mutually exclusive;
(d) the first postulate.

6.37 Postulate 3 and Postulate 2.   6.39 Postulate 1.

6.41 If Q and R are mutually exclusive, then $P(Q' \cap R')$ would equal 0.43; therefore, Q and R are not mutually exclusive.

6.43 (a) $1 - 0.62 = 0.38$; (b) $1 - 0.28 - 0.62 = 0.10$;
(c) $0.28 + 0.62 = 0.90$.

6.45 (a) 0.998; (b) 0.007;
(c) 0.993.

6.47 (a) The odds are $\frac{6}{10}$ to $\frac{4}{10}$, or 3 to 2.
(b) The odds are $\frac{11}{16}$ to $\frac{5}{16}$, or 11 to 5.
(c) The odds are against it $\frac{7}{9}$ to $\frac{2}{9}$, or 7 to 2.

6.49 The odds that it will not arrive on time are $\frac{9}{13}$ to $\frac{4}{13}$, or 9 to 4.

6.51 (a) The probability is $\frac{34}{34+21} = \frac{34}{55}$.
(b) The probability is $\frac{11}{11+2} = \frac{11}{13}$ that they will not all be $1 bills.
(c) The probability is $\frac{5}{5+1} = \frac{5}{6}$ that we will not get a meaningful word.

6.53 Let p be the subjective probability she assigns to her team winning the game.
(a) $12p = 4(1-p)$, so that $16p = 4$ and $p = \frac{1}{4}$;
(b) $12p < 4(1-p)$, so that $16p < 4$ and $p < \frac{1}{4}$.

6.55 He is claiming that the probability that he will run for the House is $\frac{1}{3}$, the probability that he is running for the Senate is $\frac{1}{5}$, and the probability that he is running for one or the other is $\frac{7}{12}$. Since running for the House and running for the Senate are mutually exclusive, we ask if $\frac{1}{3}+\frac{1}{5}$ is equal to $\frac{7}{12}$. Since it is not equal, his probabilities are inconsistent.

6.57 The probability that the first Porsche will win is $\frac{1}{5}$ and the probability that the second will win is $\frac{1}{6}$. The probability that either will win is then $\frac{11}{30}$, so that he should assign odds of 19 to 11 against either of them winning.

6.59 The event in (b) has probability $\frac{1}{1{,}024}$, which is small but certainly not zero. Since commercial airline crashes in the United States occur several times a year, there is a very small, but non-zero, probability that three will occur on the same day. The events in (a) and (c) have incredibly small probabilities; you could reasonably assign them the value zero.

6.61 Bets (a) and (b) are attractive.

6.63 (a) $0.29+0.23=0.52$; (b) $0.16+0.20+0.12+0.23=0.71$;
(c) $0.16+0.20+0.12=0.48$.

6.65 (a) $0.02+0.12+0.21+0.29=0.64$;
(b) $1-0.02=0.98$;
(c) $0.21+0.29+0.24+0.08=0.82$;
(d) $0.08+0.03+0.01=0.12$.

6.67 (a) $\frac{30}{45}=\frac{2}{3}$; (b) $10\times\frac{1}{45}=\frac{2}{9}$;
(c) $\frac{6}{45}=\frac{2}{15}$; (d) $\frac{6}{45}=\frac{2}{15}$.

6.69 (a) $0.15+0.13+0.17+0.10=0.55$; (b) $0.15+0.13+0.10+0.15=0.53$;
(c) $0.15+0.08=0.23$; (d) $1-(0.12+0.08)=0.80$.

6.71 $0.88+0.62-0.55=0.95$.

6.73 (a) $0.28+0.11-0.04=0.35$; (b) $1-0.35=0.65$.

6.75 (a) $0.37-0.10=0.27$; (b) $0.37+0.13-0.10=0.40$;
(c) $1-0.40=0.60$.

6.77 (a) There must be a mistake because $0.84+0.71-0.53=1.02>1$.
(b) There must be a mistake because $0.48+0.36+0.12=0.96<1$.

6.79 (a) $P(Q|W)$; (b) $P(W'|Q)$; (c) $P(Q'|W')$.

6.81 (a) The probability that a worker who meets the production quota is well trained.
(b) The probability that a worker who is well trained will not meet the production quota.
(c) The probability that a worker who does not meet the production quota is not well trained.

6.83 (a) $P(E|G)$; (b) $P(G'|A')$; (c) $P(A|E \cap G)$.

6.85 (a) The probability that an applicant who is employed will have the application approved.
(b) The probability that an applicant who is not employed will not have the application approved.
(c) The probability that an applicant who has the application approved will not be employed or not have a good credit rating.

6.87 (a) $\dfrac{36}{80}$; (b) $\dfrac{20}{80}$; (c) $\dfrac{24}{80}$; (d) $\dfrac{8}{80}$;
(e) $\dfrac{24}{36}$; (f) $\dfrac{8}{20}$; (g) $\dfrac{8}{44}$; (h) $\dfrac{72}{80}$.

6.89 (a) $\dfrac{24/80}{36/80} = \dfrac{24}{36}$; (b) $\dfrac{8/80}{20/80} = \dfrac{8}{20}$.

6.91 (a) $\dfrac{0.18}{0.45} = \dfrac{2}{5}$; (b) $\dfrac{0.18}{0.36} = \dfrac{1}{2}$.

6.93 (a) $\dfrac{0.72}{0.80} = 0.90$; (b) $\dfrac{0.03}{0.20} = 0.15$.

6.95 $(0.40)(0.75) = 0.30$.

6.97 (a) $\dfrac{1}{4} \cdot \dfrac{1}{4} = \dfrac{1}{16}$; (b) $\dfrac{1}{4} \cdot \dfrac{12}{51} = \dfrac{1}{17}$.

6.99 (a) $\dfrac{12}{30} \cdot \dfrac{11}{29} = \dfrac{22}{145}$; (b) $\dfrac{12}{30} \cdot \dfrac{12}{30} = \dfrac{4}{25}$.

6.101 Since $P(A \cap B) \le P(A)$ and $P(A) = 0$, we decide that $P(A \cap B) = 0$. Then $P(A|B) = \dfrac{P(A \cap B)}{P(B)} = 0$, so that $P(A|B) = P(A)$.

6.103 Since $(0.80)(0.95) = 0.76$, events A and C are independent.

6.105 $P(M) = \dfrac{3}{8}$, $P(N) = \dfrac{2}{3}$, and $P(M \cap N) = \dfrac{1}{5}$. Since $\dfrac{3}{8} \cdot \dfrac{2}{3} = \dfrac{1}{4} \ne \dfrac{1}{5}$, events M and N are not independent.

6.107 $\dfrac{1}{2} \cdot \dfrac{1}{2} \cdot \dfrac{1}{2} \cdot \dfrac{1}{2} \cdot \dfrac{1}{2} \cdot \dfrac{1}{2} \cdot \dfrac{1}{2} \cdot \dfrac{1}{2} = \dfrac{1}{256}$.

6.109 (a) $P(X|\text{Red, Deck A}) = \dfrac{5}{10} > P(X|\text{Green, Deck A}) = \dfrac{14}{30}$.

(b) $P(X|\text{Red, Deck B}) = \dfrac{5}{25} > P(X|\text{Green, Deck B}) = \dfrac{2}{15}$.

(c) The combined deck has this breakdown

	Red	Green
X	10	16
Y	25	29

(d) For the combined deck
$P(X|\text{Red, Combined Deck}) = \dfrac{10}{35} < P(X|\text{Green, Combined}) = \dfrac{16}{45}$.

6.111 $\dfrac{5}{10} \cdot \dfrac{4}{9} \cdot \dfrac{3}{8} = \dfrac{1}{12}$.

6.113 $(0.6)(0.8)(0.2)(0.3) = 0.0288$.

6.115 (a) $(0.4)(0.82) + (0.96)(0.03) = 0.0616$; (b) $\dfrac{(0.04)(0.82)}{(0.0616)} \approx 0.5325$.

6.117 (a) $(0.8)(0.14) + (0.2)(0.65) = 0.242$; (b) $\dfrac{(0.2)(0.65)}{(0.242)} \approx 0.5372$.

6.119 $\dfrac{(0.08)(0.95)}{(0.08)(0.95) + (0.92)(0.02)} = \dfrac{0.0760}{0.0944} \approx 0.8051$.

6.121 Let D denote the event that a person has the disease, and let T denote the event that a person is identified by the test having the disease.

(a) $P(D|T) = \dfrac{P(D \cap T)}{P(T)} = \dfrac{P(T|D) \cdot P(D)}{P(T|D) \cdot P(D) + P(T|D') \cdot P(D')}$

$= \dfrac{0.96 \cdot 0.02}{0.96 \cdot 0.02 + 0.06 \cdot 0.98} = \dfrac{0.0192}{0.0192 + 0.0588} \approx 0.246$;

(b) $P(D'|T') = \dfrac{P(D' \cap T')}{P(T')} = \dfrac{P(T'|D') \cdot P(D')}{P(T'|D) \cdot P(D) + P(T'|D') \cdot P(D')}$

$= \dfrac{0.94 \cdot 0.98}{0.04 \cdot 0.02 + 0.94 \cdot 0.98} = \dfrac{0.9212}{0.0008 + 0.9212} \approx 0.999$;

Because the disease is rare, getting a positive test is only a weak suggestion (probability 0.246) that a person has the disease. Getting a negative test virtually guarantees (probability 0.999) that a person does not have the disease.

6.123 (a) $(0.48)(0.90) + (0.38)(0.15) + (0.14)(0.80) = 0.601$;

(b) $\dfrac{(0.38)(0.85)}{1 - 0.601} \approx 0.81$.

6.125 Let E denote the event that there is an explosion. Denote the four possible causes as C_1, C_2, C_3, C_4; we are given $P(C_1) = 0.30$, $P(C_2) = 0.40$, $P(C_3) = 0.15$, and $P(C_4) = 0.15$. We can now compute:

$$P(E) = P(E|C_1) \cdot P(C_1) + P(E|C_2) \cdot P(C_2) + P(E|C_3) \cdot P(C_3) + P(E|C_4) \cdot P(C_4)$$
$$= 0.25(0.30) + 0.20(0.40) + 0.40(0.15) + 0.75(0.15)$$
$$= 0.3275.$$

Now we have

$$P(C_1|E) = \frac{P(E|C_1) \cdot P(C_1)}{P(E)} = \frac{0.25(0.30)}{0.3275} \approx 0.229,$$

$$P(C_2|E) = \frac{P(E|C_2) \cdot P(C_2)}{P(E)} = \frac{0.20(0.40)}{0.3275} \approx 0.244,$$

$$P(C_3|E) = \frac{P(E|C_3) \cdot P(C_3)}{P(E)} = \frac{0.40(0.15)}{0.3275} \approx 0.183, \text{ and}$$

$$P(C_4|E) = \frac{P(E|C_4) \cdot P(C_4)}{P(E)} = \frac{0.75(0.15)}{0.3275} \approx 0.344.$$

Based on these figures, the most likely cause of the explosion is C_4, purposeful action. It should be emphasized that this conclusion depends on the subjective probability estimates.

6.127 If $2P$ denotes the box with two pennies, PD the box with a penny and a dime, $2D$ the box with two dimes, and P the event that the coin we pick is a penny, we get the situation pictured in the following diagram.

```
           2P ——1—— P
      1/3 /
         / 1/3
         <——— PD ——1/2—— P
         \
      1/3 \
           2D ——0—— P
```

Thus, the probability that the other coin is also a penny, namely, that it came from $2P$, is

$$\frac{\left(\frac{1}{3}\right) \cdot 1}{\frac{1}{3} \cdot 1 + \frac{1}{3} \cdot \frac{1}{2} + \frac{1}{3} \cdot 0} = \frac{2}{3}.$$

6.129

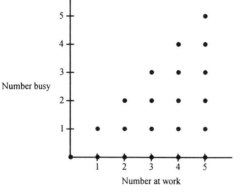

Number busy (vertical axis), Number at work (horizontal axis)

6.131 (a) $(0.2)(0.14) + (0.4)(0.04) + (0.4)(0.08) = 0.076$;

(b) $\dfrac{(0.4)(0.08)}{0.076} = \dfrac{0.032}{0.076} \approx 0.421.$

6.133 $(0.26)^4 \approx 0.0046.$

6.135 Since $(0.55)(0.18) = 0.099$, the two events are independent.

6.137 (a) The probability is $\dfrac{5}{5+19} = \dfrac{5}{24}$ that the horse will win.

(b) The probability is $\dfrac{13}{13+3} = \dfrac{13}{16}$ that at most three of the cards will be red.

(c) The probability is $\dfrac{40}{40+37} = \dfrac{40}{77}$ that one man and one woman will be selected.

6.139 The probability is $\dfrac{17}{17+8} = \dfrac{17}{25} = 0.68$.

6.141 (a) $0.01 + 0.02 + 0.05 + 0.14 + 0.16 = 0.38$;
(b) $0.18 + 0.15 + 0.09 = 0.42$;
(c) $0.14 + 0.16 + 0.20 = 0.50$.

6.143 Joe feels that the probability is 0.636 favoring the 49ers, and Walter feels that this probability is 0.714. (Both figures were rounded to three decimals.) Any bet for which the odds favoring the 49ers translates to a probability between 0.636 and 0.714 is plausible. Such a bet is 2 to 1 favoring the 49ers, for which the corresponding probability is 0.667. In such a bet Joe would put $10 on the Rams and Walter would put up $20 on the 49ers.

6.145 Let n be the number of seeds planted; $1-(1-0.025)^n \geq 0.90$, $(0.975)^n \leq 0.10$, or $n \geq 91$. Hence 91 is the smallest number of seeds which should be planted in order that the probability of getting at least one plant be 0.90 or more.

6.147 (a) $\dfrac{282,000}{375,000} = \dfrac{94}{125}$; (b) $\dfrac{300,000}{375,000} = \dfrac{4}{5}$; (c) $\dfrac{240,000}{300,000} = \dfrac{4}{5}$;

(d) $\dfrac{240,000}{282,000} = \dfrac{40}{47}$; (e) $\dfrac{33,000}{75,000} = \dfrac{11}{25}$; (f) $\dfrac{60,000}{93,000} = \dfrac{20}{31}$.

6.149 $\dfrac{28}{40} \cdot \dfrac{27}{39} \cdot \dfrac{26}{38} \cdot \dfrac{25}{37} \approx 0.224$.

6.151 The probability that the movie will be rated G is $\dfrac{1}{1+8} = \dfrac{1}{9}$, the probability that it will be rated PG is $\dfrac{3}{3+15} = \dfrac{1}{6}$, and the probability that it will be rated G or PG is $\dfrac{5}{5+13} = \dfrac{5}{18}$. Since $\dfrac{1}{9} + \dfrac{1}{6} = \dfrac{5}{18}$, the probabilities are consistent.

6.153 Let R denote that even that the person to be tested is a security risk; we are given $P(R) = 0.02$. Let F denote the event that this person fails the lie detector test; we are given $P(F'|R) = 0.10$ and $P(F|R') = 0.08$. We note that
$P(F) = P(F|R) \cdot P(R) + P(F|R') \cdot P(R') = 0.90(0.02) + 0.08(0.98) = 0.0964$.

(a) $P(R|F) = \dfrac{P(F|R) \cdot P(R)}{P(F)} = \dfrac{0.90(0.02)}{0.0964} \approx 0.187$.

(b) $P(R'|F') = \dfrac{P(F'|R') \cdot P(R')}{P(F')} = \dfrac{0.92(0.98)}{0.9036} \approx 0.998$.

36   Chapter 6 *Some Rules of Probability*

6.155   (a) $\left(\dfrac{1}{2}\right)^{12} = \dfrac{1}{4,096}$;   (b) $\left(\dfrac{1}{2}\right)^{12} = \dfrac{1}{4,096}$.

6.157   (a)   The person will visit Portugal, but neither Belgium nor England.
          (b)   The person will visit Belgium and England.
          (c)   The person will visit Belgium, but not England.
          (d)   The person will visit Belgium and/or Portugal, but not England.
          (e)   The person will visit neither Belgium nor England.

6.159   (a) $(0.3)(0.02) + (0.3)(0.25) + (0.2)(0.10) + (0.2)(0.05) = 0.0435$;
          (b) $\dfrac{(0.03)(0.025) + (0.2)(0.10)}{0.0435} = \dfrac{0.0275}{0.0435} \approx 0.632$.

6.161   (a)   The death rates at hospital A are 0.017, 0.060, and 0.333 for the three risk categories. The death rates at hospital B are 0.000, 0.044, and 0.225 for the three risk categories. The death rate is higher at hospital A for each risk category.
          (b)   Hospital A has a total of 28 deaths in 280 operations, and its overall death rate is 0.100. Hospital B has a total of 40 deaths in 290 operations, and an overall death rate is 0.138. Hospital B has a higher overall death rate, in spite of having a better rate at each specific rate category.
          (c)   About 20% of the cases in Hospital A are high risk, whereas more than 50% of the cases in Hospital B are high risk.

CHAPTER 7

Expectations and Decisions

7.1 $120 \cdot \dfrac{1}{600} = \$0.20$.

7.3 $1.00 - 10.00 \cdot \dfrac{1}{13} \approx \$0.23$.

7.5 (a) The expectation is $0.5 \times \$10,000 = \$5,000$ for each player.
(b) The player whose probability of winning is 0.65 has an expectation of $0.65 \times (\$10,000) = \$6,500$ and the opponent has an expectation of $\$3,500$.
(c) The player whose probability of winning is 0.85 has an expectation of $0.85 \times (\$10,000) = \$8,500$ and the opponent has an expectation of $\$1,500$.

7.7 Since the two expectations are $5.00 \cdot \dfrac{1}{20} = \$0.25$ and $14.00 \cdot \dfrac{1}{50} = \$0.28$, it would be smarter to draw a slip of paper from box 2.

7.9 $4 \cdot \dfrac{1}{8} + 5 \cdot \dfrac{1}{4} + 6 \cdot \dfrac{5}{16} + 7 \cdot \dfrac{5}{16} = 5\dfrac{13}{16}$.

7.11 $1.00 \cdot \dfrac{3}{4} + 1.40 \cdot \dfrac{3}{20} + 2.00 \cdot \dfrac{1}{10} = \$1.16$.

7.13 $0(0.12) + 1(0.25) + 2(0.39) + 3(0.18) + 4(0.06) = 1.81$ burglaries.

7.15 (a) If p is her probability of winning, $p < \dfrac{1}{4}$.
(b) $p > 0.15$.

7.17 If p is his subjective probability about his chances of winning twenty games, then the expected return of the salary-plus-bonus choice is $\$1,800,000 + \$350,000 p$. If he feels that this is less than $\$2,000,000$, he believes that $p < 0.571$.

7.19 If D is the cash value he attaches to owning the drill, we have $26 = D \cdot \dfrac{1}{4} + 5 \cdot \dfrac{3}{4}$, $104 = D + 15$, and $D = \$89$.

7.21 Since the expectations is $1 \cdot \dfrac{1}{4} + 3 \cdot \dfrac{1}{4} + 5 \cdot \dfrac{1}{2} = \$3.50$, it is not rational to pay $\$4$.

7.23 Since the expectation is $10,800(0.40) - 7,000(0.60) = \$120$, it is not worth her time.

7.25 (a) If the driver goes to the shopping center first, the expected distance is $\frac{2}{3}(22+22)+\frac{1}{3}(22+8+18) = 45\frac{1}{3}$ miles. If the driver goes to the barn first, the expected distance is $\frac{1}{3}(18+18)+\frac{2}{3}(18+8+22) = 44$ miles. So, his preferred strategy is to go to the barn first.

(b) If the driver goes to the shopping center first, the expected distance is $\frac{3}{4}(22+22)+\frac{1}{4}(22+8+18) = 45$ miles. If the driver goes to the barn first, the expected distance is $\frac{3}{4}(18+8+22)+\frac{1}{4}(18+18) = 45$ miles. So, the two choices are equivalent in terms of expected distances.

7.27 Since the expected profits are $4,500,000 \cdot \frac{1}{3} - 2,700,000 \cdot \frac{2}{3} = -\$300,000$, and $-1,800,000 \cdot \frac{1}{3} + 450,000 \cdot \frac{2}{3} = -\$300,000$, it does not matter whether they continue the operation.

7.29 The expected profits are $328,000(0.40) - 120,000(0.60) = \$59,200$ and $160,000(0.40) - 16,000(0.60) = \$73,600$. Delaying the expansion maximizes the expected profit.

7.31 (a) He should expand now, since this is the only strategy that has a chance of making $328,000.

(b) If the delicatessen owner is certain that the demand will be limited to 4 pies, then his maximax procedure is to stock 4 pies.

7.33 (a) The expected distance with perfect information is $36 \cdot \frac{1}{6} + 44 \cdot \frac{5}{6} = 42\frac{2}{3}$ miles and the expected value of perfect information is $44\frac{2}{3} - 42\frac{2}{3} = 2$ miles.

(b) The expected profit with perfect information is $4,500,000 \cdot \frac{1}{2} + 450,000 \cdot \frac{1}{2} = \$2,475,000$ and the expected value of perfect information is $2,475,000 - 900,000 = \$1,575,000$. It would be worthwhile to spend the $500,000.

7.35 (a) $300 \cdot \frac{2}{5} + 280 \cdot \frac{1}{5} - 20 \cdot \frac{1}{5} - 200 \cdot \frac{1}{5} = \$132$;

(b) $280 \cdot \frac{2}{5} + 300 \cdot \frac{1}{5} + 120 \cdot \frac{1}{5} - 20 \cdot \frac{1}{5} = \$192$.

7.37 The midrange, which is $\frac{\$1200 + \$2800}{2} = \$2,000$.

7.39   Suppose that he buys M papers at the beginning of the day, where M is an integer in the set $\{1, 2, \cdots, 100\}$. The cost is $15M$ cents. The probability is:
0.01 that he will sell one paper, for a net profit of $40 - 15M$;
0.01 that he will sell two papers, for a net profit of $80 - 15M$;
0.01 that he will sell M papers, for a net profit of $40M - 15M$;
$1 - 0.01M$ that the demand will exceed M papers, for a net profit of $40M - 15M$.
The mathematics expectation is found to be
$0.01(40 - 15M) + 0.01(80 - 15M) + \cdots + 0.01(40M - 15M) + (1 - 0.01M)(40M - 15M)$
$= 0.01(40 + 80 + \cdots + 40M) + (1 - 0.01M)(40M)$
$\quad + (0.01 + 0.01 + \cdots + 0.01 + (1 - 0.01M))(-15M)$
$= 0.01(40)(1 + 2 + \cdots + M) + (1 - 0.01M)(40M) - 15M$
$= 0.01(40)\dfrac{M(M+1)}{2} + (1 - 0.01M)(40M) - 15M$
$= -0.2M^2 + 25.2M$.

The value of M which maximizes this is $M = 63$, for which the maximum expected profit is 793.8 cents, $7.938. This maximizing value of M can be found by trial-and-error (which works well enough when searching for an integer), or somewhat more quickly by calculus.

7.41   If p is the probability that the play will be a success, then $150{,}000 > 250{,}000p + 0(1-p)$ and $p < \dfrac{150{,}000}{250{,}000} = 0.60$.

7.43   If the value Ms. McCall places on the hotel stay is x, $\dfrac{\$500 + x}{1{,}200} = \$2$ and $x = \$1{,}900$.

7.45   If U is the utility she attaches to the bottle of perfume, then $2 \cdot \dfrac{1}{4} + U \cdot \dfrac{3}{4} = 20$, $2 + 3U = 80$, $3U = 78$, and $U = \dfrac{78}{3} = \$26.00$.

7.47   The probability is at least $\dfrac{2}{2+1} = \dfrac{2}{3}$, but less than $\dfrac{3}{3+1} = \dfrac{3}{4}$.

7.49   (a)   If p is the probability that a person will ask for double his money back, then $1.00 > 1.25(1 - p) - 1.25p$, $1.00 > 1.25 - 2.50p$, $2.50p > 0.25$, and $p > \dfrac{0.25}{2.50} = 0.10$.
   (b)   Reversing the inequality sign, we get $p < 0.10$.
   (c)   Replacing the inequality sign with an equal sign, we get $p = 0.10$.

7.51   Since the expected cost with perfect information is $80.00(0.75) + 72.80(0.25) = \$78.20$, the expected value of perfect information is $81.80 - 78.20 = \$3.60$ and it is worthwhile to spend the \$2.40 on a long-distance call.

7.53 If she makes the reservation at hotel *A*, the minimum is $80.00, and if she makes the reservation at hotel *B*, the minimum is $72.80. Since $72.80 is the smaller of the two, she should make her reservation at hotel *B*.

7.55 The midrange, which is $\dfrac{30+33}{2} = 31.5$ miles per gallon.

CHAPTER 8

Probability Distributions

8.1 (a) No, since $f(1)+f(2)+f(3) = 0.42+0.31+0.37 = 1.10 > 1$.
(b) No, since $f(3) > 1$.
(c) Yes, since none of the probabilities is negative and $\frac{10}{33}+\frac{1}{3}+\frac{12}{33} = 1$.

8.3 (a) Yes, since none of the probabilities is negative and
$$f(1)+f(2)+\cdots+f(10) = \frac{1}{10}+\frac{1}{10}+\cdots+\frac{1}{10} = 1.$$
(b) No, since $f(10)+f(1)+\cdots+f(8) = \frac{9}{8} > 1$.
(c) Yes, since none of the probabilities is negative and
$$f(1)+f(2)+f(3)+f(4)+f(5) = \frac{2}{20}+\frac{3}{20}+\frac{4}{20}+\frac{5}{20}+\frac{6}{20} = 1.$$
(d) No, since $f(0)+f(1)+f(2)+f(3) = \frac{1}{15}+\frac{3}{15}+\frac{5}{15}+\frac{7}{15} = \frac{16}{15} > 1$.

8.5 $\binom{6}{4} 0.60^4 \, 0.40^2 \approx 0.311$.

8.7 $\binom{6}{4}(0.55)^4(0.45)^2 \approx 0.278$.

8.9 $\binom{10}{3}(0.2)^3(0.8)^7 \approx 0.201$.

8.11 (a) $\binom{10}{4}(0.40)^4(0.60)^6 \approx 0.251$; (b) 0.251.

8.13 (a) $0.002+0.009+0.032+0.086+0.172 = 0.301$;
(b) $0.086+0.172+0.250+0.250+0.154+0.044 = 0.956$;
(c) 0.250;
(d) $0.032+0.086+0.172+0.250+0.250 = 0.790$.

8.15 (a) 0.397;
(b) $0.168 + 0.376 + 0.397 = 0.941$;
(c) $0.001 + 0.009 = 0.010$.

8.17 (a) $0.001 + 0.009 + 0.046 + 0.147 = 0.203$;
(b) $0.209 + 0.090 + 0.017 = 0.316$;
(c) $0.041 + 0.008 + 0.001 = 0.050$.

8.19 The probabilities are 0.010, 0.060, 0.161, 0.251, 0.251, 0.167, 0.074, 0.021, 0.004, and 0.000.

8.21 (a) 0.311; (b) 0.230; (c) 0.654.

```
MTB > Set C1
DATA> 1(0 : 14 / 1)1
DATA> End.
MTB > PDF C1;
SUBC>   Binomial 14 .24.
```

Probability Density Function

Binomial with n = 14 and p = 0.240000

x	P(X = x)	x	P(X = x)
0.00	0.0214	8.00	0.0064
1.00	0.0948	9.00	0.0013
2.00	0.1946	10.00	0.0002
3.00	0.2459	11.00	0.0000
4.00	0.2135	12.00	0.0000
5.00	0.1348	13.00	0.0000
6.00	0.0639	14.00	0.0000
7.00	0.0231		

(a) $0.0214 + 0.0948 + 0.1946 = 0.3108$;
(b) $0.1348 + 0.0639 + .0231 + .0064 + .0013 + .0002 + 0.0000 = 0.2297$;
(c) $0.1946 + 0.2459 + 0.2135 = 0.6540$.

8.23 (a) $\dfrac{\binom{n}{x+1}p^{x+1}(1-p)^{n-x-1}}{\binom{n}{p}p^x(1-p)^{n-x}} = \dfrac{n!x!(n-x)!}{(x+1)!(n-x-1)!n!} \cdot \dfrac{p}{1-p} = \dfrac{n-x}{x+1} \cdot \dfrac{p}{1-p}$;

(b) $f(0) = \binom{6}{0}(0.25)^0(0.75)^6 = \dfrac{729}{4,096}$; $f(1) = \dfrac{729}{4,096} \cdot \dfrac{6}{1} \cdot \dfrac{0.25}{0.75} = \dfrac{1,458}{4,096}$;

$f(2) = \dfrac{1,458}{4,096} \cdot \dfrac{5}{2} \cdot \dfrac{0.25}{0.75} = \dfrac{1,215}{4,096}$; $f(3) = \dfrac{1,215}{4,096} \cdot \dfrac{4}{3} \cdot \dfrac{0.25}{0.75} = \dfrac{540}{4,096}$;

$f(4) = \dfrac{540}{4,096} \cdot \dfrac{3}{4} \cdot \dfrac{0.25}{0.75} = \dfrac{135}{4,096}$; $f(5) = \dfrac{135}{4,096} \cdot \dfrac{2}{5} \cdot \dfrac{0.25}{0.75} = \dfrac{18}{4,096}$; and

$f(6) = \dfrac{18}{4,096} \cdot \dfrac{1}{6} \cdot \dfrac{0.25}{0.75} = \dfrac{1}{4,096}$.

8.25 (a) $(0.40)(0.60)^3 = 0.0864$;
(b) $(0.25)(0.75)^4 = 0.0791$;
(c) $(0.70)(0.30)^2 = 0.0630$.

8.27 (a) The probability is 0.2969, so that 15 is not enough.
(b) The probability is 0.8867, so that 20 is not enough.
(c) The probability is 0.9998, so that 30 is more than what is needed.
(d) With 21 planted, the probability is 0.9324. This is the smallest number that he can plant in order to have the probability of 12 or more suitable ferns in excess of 0.90.

(a)
```
MTB > Set C1
DATA> 1(0 : 15 / 1)1
DATA> End.
MTB > CDF C1;
SUBC>    Binomial 15 .7.
```

Cumulative Distribution Function

Binomial with n = 15 and p = 0.700000

x	P(X <= x)		x	P(X <= x)
0.00	0.0000		8.00	0.1311
1.00	0.0000		9.00	0.2784
2.00	0.0000		10.00	0.4845
3.00	0.0001		11.00	0.7031
4.00	0.0007		12.00	0.8732
5.00	0.0037		13.00	0.9647
6.00	0.0152		14.00	0.9953
7.00	0.0500		15.00	1.0000

The probability of at 12 or more is 1.0000 - 0.7031 = 0.2969, so that 15 is not enough.

(b)
```
MTB > Set C1
DATA> 1(0 : 20 / 1)1
DATA> End.
MTB > CDF C1;
SUBC>    Binomial 20 .7.
```

Cumulative Distribution Function

Binomial with n = 20 and p = 0.700000

x	P(X <= x)		x	P(X <= x)
0.00	0.0000		11.00	0.1133
1.00	0.0000		12.00	0.2277
2.00	0.0000		13.00	0.3920
3.00	0.0000		14.00	0.5836
4.00	0.0000		15.00	0.7625
5.00	0.0000		16.00	0.8929
6.00	0.0003		17.00	0.9645
7.00	0.0013		18.00	0.9924
8.00	0.0051		19.00	0.9992
9.00	0.0171		20.00	1.0000
10.00	0.0480			

The probability of at 12 or more is 1.0000 - 0.1133 = 0.8867, so that 20 is not enough.

(c) MTB > Set C1
     DATA> 1(0 : 30 / 1)1
     DATA> End.
     MTB > CDF C1;
     SUBC>   Binomial 30 .7.

Cumulative Distribution Function

Binomial with n = 30 and p = 0.700000

x	P(X <= x)	x	P(X <= x)
0.00	0.0000	16.00	0.0401
1.00	0.0000	17.00	0.0845
2.00	0.0000	18.00	0.1593
3.00	0.0000	19.00	0.2696
4.00	0.0000	20.00	0.4112
5.00	0.0000	21.00	0.5685
6.00	0.0000	22.00	0.7186
7.00	0.0000	23.00	0.8405
8.00	0.0000	24.00	0.9234
9.00	0.0000	25.00	0.9698
10.00	0.0000	26.00	0.9907
11.00	0.0002	27.00	0.9979
12.00	0.0006	28.00	0.9997
13.00	0.0021	29.00	1.0000
14.00	0.0064	30.00	1.0000
15.00	0.0169		

The probability of 12 or more is 1.0000 - 0.0002 = 0.9998, so that 30 is more than what is needed.

(d) MTB > Set C1
     DATA> 1(0 : 21 / 1)1
     DATA> End.
     MTB > CDF C1;
     SUBC>   Binomial 21 .7.

Cumulative Distribution Function

Binomial with n = 21 and p = 0.700000

x	P(X <= x)	x	P(X <= x)
0.00	0.0000	11.00	0.0676
1.00	0.0000	12.00	0.1477
2.00	0.0000	13.00	0.2770
3.00	0.0000	14.00	0.4495
4.00	0.0000	15.00	0.6373
5.00	0.0000	16.00	0.8016
6.00	0.0001	17.00	0.9144
7.00	0.0006	18.00	0.9729
8.00	0.0024	19.00	0.9944
9.00	0.0087	20.00	0.9994
10.00	0.0264	21.00	1.0000

With 21 planted, the probability of 12 or more is 1.0000 - 0.0676 = 0.9324. This is the smallest number that he can plant in order to have the probability of 12 or more suitable ferns in excess of 0.90.

8.29 $\dfrac{\binom{5}{3}\binom{11}{0}}{\binom{16}{3}} \approx 0.018$.

8.31 (a) $\dfrac{\binom{9}{0}\binom{6}{3}}{\binom{15}{3}} = 0.044$; (b) $\dfrac{\binom{9}{1}\binom{6}{2}}{455} = 0.2967$;

(c) $\dfrac{\binom{9}{2}\binom{6}{1}}{455} = 0.4747$; (d) $\dfrac{\binom{9}{3}\binom{6}{0}}{455} \approx 0.1846$.

8.33 (a) $\dfrac{\binom{1}{0}\binom{11}{2}}{\binom{12}{2}} = \dfrac{1 \cdot 55}{66} \approx 0.833$; (b) $1 - \dfrac{\binom{3}{0}\binom{9}{2}}{66} = 1 - \dfrac{36}{66} = \dfrac{30}{36} = 0.455$;

(c) $1 - \dfrac{\binom{6}{0}\binom{6}{2}}{66} = 1 - \dfrac{1 \cdot 15}{66} = \dfrac{51}{66} \approx 0.773$.

8.35 (a) Since $n = 12 > 0.05(140 + 60) = 10$, the condition is not satisfied.
(b) Since $n = 20 < 0.05(220 + 280) = 25$, the condition is satisfied.
(c) Since $n = 30 < 0.05(250 + 390) = 32$, the condition is satisfied.
(d) Since $n = 25 > 0.05(220 + 220) = 22$, the condition is not satisfied.

8.37 (a) $\dfrac{\binom{144}{3} \cdot \binom{36}{2}}{\binom{180}{5}} = \dfrac{487{,}344 \cdot 630}{1{,}488{,}847{,}536} \approx 0.206218$.

(b) In the approximation, use $n = 5$ and $p = 0.80$. The approximating probability is
$\binom{5}{3} 0.80^3 \cdot 0.20^2 = 0.204800$.

(c) The error of approximation is $0.204800 - 0.206218 = -0.001418$. In percentage terms, this error is $-\dfrac{0.001418}{0.206218} \approx -0.69\%$.

8.39 $\dfrac{\binom{175}{9}}{\binom{200}{9}} \cdot \dfrac{25}{200-9} \approx 0.038$.

8.41 Substitute $a = 12$, $b = 28$, $x = 12$, $k = 4$ into the formula, and we have

$$\frac{12-3}{40-11} \cdot \frac{\binom{12}{3}\binom{28}{8}}{\binom{40}{11}} \approx 0.0459.$$

8.43 (a) Since $500 > 100$, and $500 \cdot 0.001 = 0.05 < 1$ the conditions are satisfied.
(b) Since $100 \cdot 0.12 = 12 > 10$, the conditions are not satisfied.
(c) Since $60 < 100$, the conditions are not satisfied.

8.45 Using $np = 3$, the approximating probability is $\dfrac{e^{-3}3^4}{4!} = 0.1680$.

8.47 $np = 200 \cdot 0.032 = 6.4$. Then

(a) $\dfrac{(6.4)^6 e^{-6.4}}{6!} \approx 0.159$;

(b) $\displaystyle\sum_{i=0}^{8} \dfrac{(64)^i e^{-6.4}}{i!} \approx 0.803$;

(c) $1 - \displaystyle\sum_{i=0}^{4} \dfrac{(6.4)^i e^{-6.4}}{i!} \approx 1 - 0.235 = 0.765.$

8.49 (a) $\dfrac{4.2^5 e^{-4.2}}{5!} \approx 0.163$; (b) $\displaystyle\sum_{i=0}^{7} \dfrac{4.2^i \cdot e^{-4.2}}{i!} \approx 0.936.$

8.51 (a) $f(2) = \dfrac{(4.4)^2 e^{-4.4}}{2!} \approx \dfrac{(19.36)(0.012)}{2} = 0.116$;

(b) $f(3) = \dfrac{(4.4)^3 (0.012)}{3!} = 0.170$;

(c) $f(0) + f(1) + f(2) + f(3) \approx \dfrac{(4.4)^0(0.012)}{0!} + \dfrac{(4.4)^1(0.012)}{1!} + \dfrac{(4.4)^2(0.012)}{2!}$

$+ \dfrac{(4.4)^3(0.012)}{3!} \approx 0.012 + 0.053 + 0.116 + 0.170 = 0.351$

8.53 (a) 0.0988; (b) $1 - 0.9966 = 0.0034.$

8.55 (a) $\dfrac{5!}{1!2!1!1!}\left(\dfrac{1}{8}\right)\left(\dfrac{3}{8}\right)^2\left(\dfrac{3}{8}\right)\left(\dfrac{1}{8}\right) \approx 0.049$;

(b) $\dfrac{6!}{1!3!2!0!}\left(\dfrac{1}{8}\right)\left(\dfrac{3}{8}\right)^3\left(\dfrac{3}{8}\right)^2\left(\dfrac{1}{8}\right)^0 \approx 0.056.$

8.57 $\dfrac{9!}{4!2!3!0!}\left(\dfrac{9}{16}\right)^4\left(\dfrac{3}{16}\right)^2\left(\dfrac{3}{16}\right)^3\left(\dfrac{1}{16}\right)^0 \approx 0.029.$

8.59 (a) $\dfrac{13!}{4!3!3!3!}\left(\dfrac{1}{4}\right)^4\left(\dfrac{1}{4}\right)^3\left(\dfrac{1}{4}\right)^3\left(\dfrac{1}{4}\right)^3 \approx 0.0179$;

(b) $4\cdot\dfrac{13!}{4!3!3!3!}\left(\dfrac{1}{4}\right)^4\left(\dfrac{1}{4}\right)^3\left(\dfrac{1}{4}\right)^3\left(\dfrac{1}{4}\right)^3 \approx 0.0716$.

8.61 (a) $0(0.20)+(0.40)+2(0.30)+3(0.10)=1.3$;
(b) $(0-1.3)^2(0.2)+(1-1.3)^2(0.4)+(2-1.3)^2(0.3)+(3-1.3)^2(0.1)=0.81$;
(c) $0^2(0.2)+1^2(0.4)+2^2(0.3)+3^2(0.1)-(1.3)^2=0.81$.

8.63 $\mu=0(0.48)+1(0.25)+2(0.14)+3(0.08)+4(0.04)+5(0.01)=0.98$, and
$\sigma^2=0^2(0.48)+1^2(0.25)+2^2(0.14)+3^2(0.08)+4^2(0.04)+5^2(0.01)-0.98^2=1.46$. Then it follows $\sigma=\sqrt{1.46}\approx 1.21$.

8.65 (a) $\mu=0(0.001)+1(0.008)+2(0.041)+3(0.124)+4(0.232)$
$+5(0.279)+6(0.209)+7(0.090)+8(0.017)=4.805$

and

$\sigma^2=0^2(0.001)+1^2(0.008)+2^2(0.041)+3^2(0.124)+4^2(0.232)$
$+5^2(0.279)+6^2(0.209)+7^2(0.090)+8^2(0.017)-4.805^2 \approx 1.909$;

(b) $\mu=8(0.60)=4.8$ and $\sigma^2=8(0.60)(0.40)=1.92$.

8.67 $\mu=20(0.80)=16$ and $\sigma=\sqrt{20(0.80)(0.20)}\approx 1.79$.

8.69 (a) $\mu=460\left(\dfrac{1}{2}\right)=230$; $\sigma=\sqrt{460\left(\tfrac{1}{2}\right)\left(1-\tfrac{1}{2}\right)}\approx 10.72$;

(b) $\mu=1{,}200\left(\dfrac{1}{6}\right)=200$; $\sigma=\sqrt{1{,}200\left(\tfrac{1}{6}\right)\left(1-\tfrac{1}{6}\right)}\approx 12.91$;

(c) $\mu=200(0.36)=72$; $\sigma=\sqrt{200(0.36)(1-0.36)}\approx 6.79$;

(d) $\mu=450(0.08)=36$; $\sigma=\sqrt{450(0.08)(1-0.08)}\approx 5.75$;

(e) $\mu=650(0.68)=442$; $\sigma=\sqrt{650(0.68)(1-0.68)}\approx 11.89$.

8.71 (a) $\mu=0(0.295)+1(0.491)+2(0.196)+3(0.018)=0.937$;

(b) $\mu=\dfrac{3\cdot 5}{16}=0.9375$.

8.73 (a) $\sigma^2=3\dfrac{9\cdot 6}{(9+6)^2}\cdot\dfrac{9+6-3}{9+6-1}\approx 0.6171$, hence $\sigma=\sqrt{0.6171}\approx 0.7856$;

(b) $\sigma^2=3\dfrac{5\cdot 11}{(5+11)^2}\cdot\dfrac{5+11-3}{5+11-1}\approx 0.5586$, hence $\sigma=\sqrt{0.5586}\approx 0.7474$.

8.75 $\mu=0(0.449)+1(0.360)+2(0.144)+3(0.038)+4(0.008)+5(0.001)=0.799$, and this is very close to $\lambda=0.8$.

8.77 $\sigma^2 = 0^2(0.449) + 1^2(0.360) + 2^2(0.144) + 3^2(0.038)$
$+ 4^2(0.008) + 5^2(0.001) - 0.799^2 \approx 0.793$.
Thus $\sigma = \sqrt{0.793} \approx 0.8905$, which is very close to $\sqrt{0.8} \approx 0.8944$.

8.79 The probability is at least $1 - \frac{1}{2^2} = 0.75$ that we'll get less than $2.4 + 2(1.549) = 5.498$ emitted in one second.

8.81 If $1 - \frac{1}{k^2} = 0.96$, then $k = 5$.

(a) Since $\mu = 10,000(0.50) = 5,000$ and $\sigma = \sqrt{10,000(0.50)(0.50)} = 50$, the probability is at least 0.96 that the number of heads will be between $5,000 - 5 \cdot 50 = 4,750$ and $5,000 + 5 \cdot 50 = 5,250$, and hence that the proportion of heads will be between $\frac{4,750}{10,000} = 0.475$ and $\frac{5,250}{10,000} = 0.525$.

(b) Since $\mu = 1,000,000(0.50) = 500,000$ and $\sigma = \sqrt{1,000,000(0.50)(0.50)} = 500$, the probability is at least 0.96 that the number of heads will be between $500,000 - 5 \cdot 500 = 497,500$ and $500,000 + 5 \cdot 500 = 502,500$, and hence that the proportion of heads will be between $\frac{497,500}{1,000,000} = 0.4975$ and $\frac{502,500}{1,000,000} = 0.5025$.

8.83 (a) 00–24, 25–74, 75–99.

8.87 (a) 0000–2465, 2466–5917, 5918–8334, 8335–9462, 9463–9857, 9858–9968, 9969–9994, and 9995–9999.

8.91 (a) The ratio is $\dfrac{\binom{30}{1}\binom{270}{11}}{\binom{300}{12}}$ to $\dfrac{\binom{30}{0}\binom{270}{12}}{\binom{300}{12}}$ which simplifies to $\dfrac{360}{259} \approx 1.39$;

(b) The ratio is $\binom{30}{1}\left(\dfrac{30}{300}\right)^1\left(1 - \dfrac{30}{300}\right)^{30-1}$ to $\binom{30}{0}\left(\dfrac{30}{300}\right)^0\left(1 - \dfrac{30}{300}\right)^{30}$, which simplifies to $\dfrac{3}{0.9} \approx 3.33$.

8.93 (a) These cannot be the values of a probability distribution, since their total is 0.98.
(b) These cannot be the values of a probability distribution, since $f(4)$ is negative.
(c) These cannot be the values of a probability distribution, since their total is 1.25.

8.95 $(0.20)(0.80)^4 \approx 0.082$.

8.97 00–45, 46–72, 73–87, 88–95, and 96–99.

8.99 (a) Since $8 < 0.05(80 + 120) = 10$, the condition is satisfied.
(b) Since $10 > 0.05(75 + 75) = 7.5$, the condition is not satisfied.
(c) Since $24 > 0.05(136 + 164) = 15$, the condition is not satisfied.

8.101 (a) 0.1985;   (b)   0.8109;   (c) $0.9684 - 0.0991 = 0.8693$.

8.103 (a) The probability is at least $1-\frac{1}{7^2}=\frac{48}{49}$ that between $1,200-7\cdot 80 = 640$ and $1,200+7\cdot 80 = 1,760$ persons will visit the tourist attraction.

(b) Since $k = \frac{1,400-1,200}{80} = \frac{1,200-1,000}{80} = 2.5$, the probability is at least $1-\frac{1}{2.5^2} = 0.84$.

8.105 (a) The hypergeometric probability is $\dfrac{\binom{1393}{48}\binom{35}{2}}{\binom{1428}{50}}$

(b) The binomial probability is $\binom{50}{2}0.0245^2 \cdot 0.9755^{48} \approx 0.223551$.

(c) The Poisson probability is $e^{-1.2255}\dfrac{1.2255^2}{2!} \approx 0.220480$.

8.107 (a) $\mu = 0(0.120)+1(0.207)+2(0.358)+3(0.162)+4(0.096)+5(0.043)+6(0.014) = 2.092$ and
$\sigma^2 = 0^2(0.120)+1^2(0.207)+2^2(0.358)+3^2(0.162)$
$\quad + 4^2(0.096)+5^2(0.043)+6^2(0.014)-(2.092)^2 \approx 1.836$.

8.109 (a) Use Table V with $n = 10$ and $p = 0.70$ to find this probability as $0.233 + 0.121 + 0.028 = 0.382$.

(b) Use Table V with $n = 10$ and $p = 0.10$ to find this probability as $0.349 + 0.387 + 0.194 = 0.930$.

(c) $\dfrac{10!}{6!3!1!}0.70^6\ 0.20^3\ 0.10^1 \approx 0.079$.

8.111 (a) 0.146;
(b) $0.146 + 0.371 + 0.440 = 0.957$;
(c) $1-(0.371+0.440) = 0.189$.

8.113 (a) $\mu = \dfrac{5\cdot 20}{20+20} = 2.5$; $\sigma = \sqrt{5\cdot \dfrac{20\cdot 20}{(20+20)^2}\cdot \dfrac{20+20-5}{20+20-1}} \approx 1.06$;

(b) $\mu = \dfrac{15\cdot 2}{2+198} = 0.15$; $\sigma = \sqrt{15\cdot \dfrac{2\cdot 198}{(2+198)^2}\cdot \dfrac{2+198-15}{2+198-1}} \approx 0.372$;

(c) $\mu = \dfrac{15\cdot 200}{20,000} = 0.15$; $\sigma = \sqrt{15\cdot \dfrac{200\cdot 19,800}{(200+19,800)^2}\cdot \dfrac{20,000-15}{20,000-1}} \approx 0.385$;

(d) $\mu = \dfrac{30\cdot 28}{28+12} = 21$; $\sigma = \sqrt{30\cdot \dfrac{28\cdot 12}{(28+12)^2}\cdot \dfrac{40-30}{40-1}} \approx 1.27$.

CHAPTER 9

The Normal Distribution

9.1 (a) The total area under the curve is $(4-1)\frac{1}{4} = \frac{3}{4} < 1$.

    (b) $f(x) < 0$ for $1 \leq x < \frac{7}{4}$.

9.3 (a) $(7-2)\frac{1}{8} = \frac{5}{8} = 0.625$;   (b) $(8.8-2.4)\frac{1}{8} = \frac{6.4}{8} = 0.8$.

9.5 $1 - \frac{1}{2} \cdot 3 \cdot \frac{3}{8} = 1 - \frac{9}{16} = \frac{7}{16}$.

9.7 (a) the first area is bigger; (b) the first area is bigger;
(c) the first area is bigger; (d) the two areas are equal;
(e) the first area is bigger; (f) the second area is bigger;
(g) the two areas are equal; (h) the first area is bigger.

9.9 (a) 0.3078; (b) 0.4515;
(c) $0.5000 - 0.1844 = 0.3156$; (d) $0.1064 + 0.5000 = 0.6064$;
(e) $0.4032 + 0.5000 = 0.9032$; (f) $0.5000 - 0.2852 = 0.2148$.

9.11 (a) $0.3212 + 0.2580 = 0.5792$; (b) $0.5 - 0.3749 = 0.1251$;
(c) $0.5 + 0.0871 = 0.5871$; (d) $0.4656 - 0.0948 = 0.3708$;
(e) $0.5 - 0.3729 = 0.1271$; (f) $0.5 + 0.2764 = 0.7764$;
(g) $0.4656 - 0.2852 = 0.1804$.

9.13 (a) $z = \pm 1.48$; (b) $z = -0.74$;
(c) $z = 1.12$; (d) $z = \pm 2.17$.

9.15 (a) $z = \pm 0.53$; (b) $z = \pm 1.56$;
(c) $z = \pm 0.44$; (d) $z = \pm 2.03$.

9.17 (a) the first probability is bigger;
(b) the second probability is bigger;
(c) the first probability is bigger;
(d) the first probability is bigger.

9.19 (a) Since $z = \frac{87.2 - 80}{4.8} = 1.50$, the probability is $0.4332 + 0.5000 = 0.9332$.

    (b) Since $z = \frac{76.4 - 80}{4.8} = -0.75$, the probability is $0.2734 + 0.5000 = 0.7734$.

    (c) Since $z = \frac{86.0 - 80}{4.8} = 1.25$ and $z = \frac{81.2 - 80}{4.8} = 0.25$, the probability is $0.3944 - 0.0987 = 0.2957$.

    (d) Since $z = \frac{88.4 - 80}{4.8} = 1.75$ and $z = \frac{71.6 - 80}{4.8} = -1.75$, the probability is $2(0.4599) = 0.9198$.

9.21 Since the entry closest to $0.5000 - 0.2000 = 0.3000$ is 0.84 corresponding to 0.2995, we get $\frac{79.2 - 62.4}{\sigma} = 0.84$ and $\sigma = \frac{16.8}{0.84} = 20$.

9.23 (a) $1 - e^{-0.4} \approx 1 - 0.670 = 0.330$;
(b) $e^{-0.5} - e^{-0.9} \approx 0.607 - 0.407 = 0.200$;
(c) $e^{-1.6} \approx 0.202$.

9.25 (a) $1 - (1 - e^{-2}) \approx 0.135$;
(b) $1 - (1 - e^{-3}) \approx 0.050$;
(c) $1 - (1 - e^{-0.5}) \approx 1 - 0.607 = 0.393$.

9.27 It cannot be used simply because we only have two points, which will always fit a line.

9.31 Roughly, $\mu = 18.6$ and $\sigma = \frac{24.2 - 13.4}{2} = 5.4$.

9.33 The data were in fact obtained from a normally distributed population, and the normal probability paper should show the points lying, at least approximately, on a straight line.

9.35 (a) 0.203; (b) 0.459; (c) 0.749.

9.37 (a) Since $z = \frac{33.4 - 38.6}{4} = -0.80$, the probability is $0.2881 + 0.5000 = 0.7881$.

(b) Since $z = \frac{34.7 - 38.6}{4} = -0.60$, the probability is $0.5000 - 0.2257 = 0.2743$.

9.39 Since the entry closest to 0.3000 is 0.2995 corresponding to $z = 0.84$, we get $\frac{x - 4.54}{0.25} = 0.84$, and then $x = 4.54 + 0.25(0.84) = 4.75$.

9.41 Since $z = \frac{6.00 - 6.02}{0.01} = -2.00$, the normal curve area is $0.5000 + 0.4772 = 0.9772$ and the percentage is 97.72%.

9.43 (a) 0.0809; (b) 0.3859; (c) 0.6950.
These answers are approximate and might reasonably be rounded back to two decimals.

9.45 Since $z = \frac{14.5 - 18.6}{3.3} \approx -1.24$ and $z = \frac{15.5 - 18.6}{3.3} \approx -0.94$, the probability is $0.3925 - 0.3264 = 0.0661$.

9.47 (a) Since $200(0.01) = 2 < 5$, the conditions are not satisfied.
(b) Since $150(0.03) = 4.5 < 5$, the conditions are not satisfied.
(c) Since $100 \cdot \frac{1}{8} = 12.5 > 5$ and $100 \cdot \frac{7}{8} = 87.5 > 5$, the conditions are satisfied.

9.49 $\mu = 20(0.5) = 10$ and $\sigma = \sqrt{20(0.5)(0.5)} \approx 2.236$. Since $z = \dfrac{12.5-10}{2.236} \approx 1.12$ and $z = \dfrac{11.5-10}{2.236} \approx 0.67$, the probability is $0.3686 - 0.2486 = 0.120$ and the error is $0.120 - 0.120134 = -0.000134$.

9.51 $\mu = 40(0.62) = 24.8$ and $\sigma = \sqrt{40(0.62)(0.338)} \approx 3.07$. Since $z = \dfrac{20.5-24.8}{3.07} \approx -1.40$, the probability is $0.5000 - 0.4129 = 0.0808$.

9.53 $\mu = 50(0.22) = 11$ and $\sigma = \sqrt{50(0.22)(0.78)} \approx 2.93$. Since $z = \dfrac{12.5-11}{2.93} \approx 0.51$, the probability is $0.5000 + 0.1950 = 0.6950$.

9.55 From the printout of Figure 8.4, the probability is 0.0036. $\mu = 150(0.05) = 7.5$ and $\sigma = \sqrt{150(0.05)(0.95)} \approx 2.67$. Since $z = \dfrac{0.5-7.5}{2.67} \approx -2.62$ and $z = \dfrac{1.5-7.5}{2.67} \approx -2.25$, the probability is $0.4956 - 0.4878 = 0.0078$. The error is $0.0078 - 0.0036 = 0.0042$ and the percentage error is $\dfrac{0.0042}{0.0036} \cdot 100\% \approx 117\%$.

9.57 The probability is $1.0000 - 0.02744 = 0.97256$. The error is $0.97256 - 0.9750 = -0.00244$, and the percentage error is $-\dfrac{0.00244}{0.9750} \times 100\% \approx -0.25\%$.

9.59 (a) Since $50 \cdot \dfrac{1}{12} \approx 4.17 < 5$, we cannot use the normal approximation; since $50 < 100$, we cannot use the Poisson approximation.
(b) Since $200(0.06) = 12 > 5$ and $200(0.94) = 188 > 5$, we can use the normal approximation; since $200(0.06) = 12 > 10$, we cannot use the Poisson approximation.
(c) Since $100(0.04) = 4 < 5$, we cannot use the normal approximation; since $100 \geq 100$ and $100(0.04) = 4 < 10$, we can use the Poisson approximation.
(d) Since $100(0.08) = 8 > 5$ and $100(0.92) = 92 > 5$, we can use the normal approximation; since $100 \geq 100$ and $100(0.08) = 8 < 10$, we can use the Poisson approximation.

9.61 Since $f(1) = \dfrac{7.5^1 e^{-7.5}}{1!} \approx 7.5(0.00055) = 0.004125$, the error is $0.004125 - 0.0036 = 0.000525$ and the percentage error is $\dfrac{0.000525}{0.0036} \cdot 100\% \approx 15\%$.

9.67 $\mu = 20(0.50) = 10$ and $\sigma = \sqrt{20(0.50)(0.50)} \approx 2.236$. Since $z = \dfrac{11.5-10}{2.236} \approx 0.67$ and $z = \dfrac{12.5-10}{2.236} \approx 1.12$, the probability is $0.3686 - 0.2486 = 0.1200$. The error is $0.1200 - 0.120 = 0$.

9.69 (a) Since $z = \dfrac{107.8 - 102.4}{3.6} \approx 1.50$, the probability is $0.5000 + 0.4332 = 0.9332$.

(b) Since $z = \dfrac{99.7 - 102.4}{3.6} = -0.75$, the probability is $0.5000 + 0.2734 = 0.7734$.

(c) Since $z = \dfrac{106.9 - 102.4}{3.6} = 1.25$ and $z = \dfrac{110.5 - 102.4}{3.6} = 2.25$, the probability is $0.4878 - 0.3944 = 0.0934$.

(d) Since $z = \dfrac{96.1 - 102.4}{3.6} = -1.75$ and $z = \dfrac{104.2 - 102.4}{3.6} = 0.50$, the probability is $0.4599 + 0.1915 = 0.6514$.

9.75 (a) Since $z = \dfrac{54.5 - 48.4}{7.5} \approx 0.81$, the probability is $0.5000 - 0.2910 = 0.2090$.

(b) Since $z = \dfrac{39.5 - 48.4}{7.5} \approx -1.19$ and $z = \dfrac{50.5 - 48.4}{7.5} \approx 0.28$, the probability is $0.3830 + 0.1103 = 0.4933$.

9.77 Since $z = \dfrac{102 - 90}{12} = 1$, the proportion of the time that job A will take longer than average job B is $0.5000 - 0.3413 = 0.1587$. Since $z = \dfrac{90 - 102}{18} \approx -0.67$, the proportion of the time that job B will take longer than the average job A is $0.5000 + 0.2486 = 0.7486$.

9.79 (a) 0.3665; (b) $2(0.2357) = 0.4714$;
(c) $0.2088 - 0.1554 = 0.0534$; (d) $0.3180 - 0.2517 = 0.1293$;
(e) $0.3888 + 0.4656 = 0.8544$.

9.81 (a) Since $z = \dfrac{26.5 - 25.3}{5.5} \approx 0.22$ and $z = \dfrac{27.5 - 25.3}{5.5} = 0.40$, the probability is $0.1554 - 0.0871 = 0.0683$.

(b) Since $z = \dfrac{26.5 - 25.3}{5.5} \approx 0.22$, the probability is $0.5000 - 0.0871 = 0.4129$.

9.83 (a) $\mu = 1{,}000(0.50) = 500$ and $\sigma = \sqrt{1{,}000(0.50)(0.50)} \approx 15.81$. Since $z = \dfrac{489.5 - 500}{15.81} \approx -0.66$, and $z = \dfrac{510.5 - 500}{15.81} \approx 0.66$, the probability is $2(0.2454) = 0.4908$.

(b) $\mu = 10{,}000(0.50) = 5{,}000$ and $\sigma = \sqrt{10{,}000(0.50)(0.50)} = 50$. Since $z = \dfrac{4{,}899.5 - 5{,}000}{50} \approx -2.01$, and $z = \dfrac{5{,}100.5 - 5{,}000}{50} \approx 2.01$, the probability is $2(0.4778) = 0.9556$.

CHAPTER 10

Sampling and Sampling Distributions

10.1 (a) $\binom{4}{2} = 6$; (b) $\binom{8}{2} = 28$; (c) $\binom{20}{2} = 190$; (d) $\binom{200}{2} = 19,900$.

10.3 (a) $\dfrac{1}{\binom{12}{4}} = \dfrac{1}{495}$; (b) $\dfrac{1}{\binom{22}{5}} = \dfrac{1}{26,334}$.

10.5
abc
abd   acd
abe   ace   ade
abf   acf   adf   aef
abg   acg   adg   aeg   afg

bcd
bce   bde
bcf   bdf   bef
bcg   bdg   beg   bfg

cde
cdf   cef
cdg   ceg   cfg

def
deg   dfg

efg

10.7 TWA and American, TWA and United, TWA and Northwest, TWA and Delta, TWA and U.S. Air, American and United, American and Northwest, American and Delta, American and U.S. Air, United and Northwest, United and Delta, United and U.S. Air, Northwest and Delta, Northwest and U.S. Air, and Delta and U.S. Air.

(a) $\dfrac{1}{15}$; (b) $\dfrac{\binom{5}{1}}{\binom{6}{2}} = \dfrac{5}{15} = \dfrac{1}{3}$.

10.9 The selected homes will be

3332	7255	6441	1561	7740
5041	3962	4044	2153	3206
5845	0111	2106	5895	6326
3281	6571	0792	0772	4732

10.11 This probability is $\frac{39}{40} \cdot \frac{38}{40} \cdot \frac{37}{40} = 0.8568$.

10.13 (a) The number of different results for three numbers drawn in succession is $100 \cdot 99 \cdot 98$. Of these, there are $6 = 3!$ patterns in which a particular three items can arise. The probability of any particular three items is then $\frac{6}{100 \cdot 99 \cdot 98} = \frac{1}{\binom{100}{3}}$.

(b) Use the same logic as in (a), but with general N rather than 100.

10.15 (a) $\frac{1}{10}$; (b) $\frac{1}{10}$; (c) $\frac{\binom{48}{3}}{\binom{50}{5}} \approx 0.0082$; (d) less than 0.01.

10.17 (a) Combine the data for January and July, those for February and August, those for March and September, those for April and October, those for May and November, and those for June and December.

(b) The sixth sample contains all the high figures for December.

10.19 (a) Here are the samples of size 2, along with their means:

Sample		Mean	Sample		Mean
115	125	120	125	205	165
115	135	125	135	185	160
115	185	150	135	195	165
115	195	155	135	205	170
115	205	160	185	195	190
125	135	130	185	205	195
125	185	155	195	205	200
125	195	160			

The probability that the sample mean will differ from the population mean 160 by more than 5 is $\frac{8}{15}$.

(b) Here are the stratified samples, one for each sex, along with their means:

Sample F	Sample M	Mean	Sample F	Sample M	Mean
115	185	150	125	205	165
115	195	155	135	185	160
115	205	160	135	195	165
125	185	155	135	205	170
125	195	160			

The probability that the sample mean will differ from the population mean 160 by more than 5 is $\frac{2}{9}$.

(c) Here are the stratified samples, one from each age group, along with their means:

Sample Y	Sample O	Mean	Sample Y	Sample O	Mean
125	115	120	135	205	170
125	195	160	185	115	150
125	205	165	185	195	190
135	115	125	185	205	195
135	195	165			

The probability that the sample mean will differ from the population mean 160 by more than 5 is $\frac{2}{3}$.

(d) The results of the relevant stratification in (b) greatly improved on the simple random sampling of (a). The stratification proposed in (d) was actually destructive for the task of estimating the mean weight.

10.21 (a) $\binom{120}{2}\binom{80}{2}\binom{40}{2} = 17{,}598{,}672{,}000$; (b) $\binom{120}{3}\binom{80}{2}\binom{40}{1} = 35{,}498{,}176$.

10.23 $n_1 = \frac{500}{2{,}000} \cdot 80 = 20$, $n_2 = \frac{1{,}200}{2{,}000} \cdot 80 = 48$, $n_3 = \frac{200}{2{,}000} \cdot 80 = 8$, and $n_4 = \frac{100}{2{,}000} \cdot 80 = 4$.

10.25 115 and 125, 115 and 135, 125 and 135, 185 and 195, 185 and 205, and 195 and 205.
The means are 120, 125, 130, 190, 195, and 200.
Since all the values are less than 155 or greater than 165, the probability is 1.
Since $\frac{2}{9} < \frac{8}{15} < 1$, stratified sampling is better than simple random sampling, which in turn is better than cluster sampling.

10.27 $n_1 = \frac{100 \cdot 10{,}000 \cdot 45}{10{,}000 \cdot 45 + 30{,}000 \cdot 60} = 20$ and $n_2 = \frac{100 \cdot 30{,}000 \cdot 60}{10{,}000 \cdot 45 + 30{,}000 \cdot 60} = 80$.

10.29 Since $\bar{x}_i = \dfrac{\sum x_i}{n_i}$, we get

$$\bar{x}_w = \dfrac{1}{N}\left(N_1 \cdot \dfrac{\sum x_1}{n_1} + N_2 \dfrac{\sum x_2}{n_2} + \cdots + N_k \cdot \dfrac{\sum x_k}{n_k}\right)$$

$$= \dfrac{1}{N}\left(\dfrac{N}{n} \cdot \sum x_1 + \dfrac{N}{n} \cdot \sum x_2 + \cdots + \dfrac{N}{n} \cdot \sum x_k\right) = \dfrac{1}{n}\left(\sum x_1 + \sum x_2 + \cdots + \sum x_k\right)$$

$$= \dfrac{1}{n}\left(\sum x\right) = \bar{x}$$

10.31 (a) 10 and 10, 10 and 10, 10 and 10, 10 and 10,
10 and 10, 10 and 10, 10 and 10, 10 and 10,
10 and 10, 10 and 10, 10 and 12, 10 and 18,
10 and 40, 10 and 12, 10 and 18, 10 and 40,
10 and 12, 10 and 18, 10 and 40, 10 and 12,
10 and 18, 10 and 40, 10 and 12, 10 and 18,
10 and 40, 12 and 18, 12 and 40, 18 and 40.

(b) They are, in the same order as in (a),
10, 10, 10, 10, 10, 10, 10, 10, 10, 10,
11, 14, 25, 11, 14, 25, 11, 14, 25, 11,
14, 25, 11, 14, 25, 15, 26, 29.

(c)

Mean	Probability
10	$\dfrac{10}{28}$
11	$\dfrac{5}{28}$
14	$\dfrac{5}{28}$
15	$\dfrac{1}{28}$
25	$\dfrac{5}{28}$
26	$\dfrac{1}{28}$
29	$\dfrac{1}{28}$

(d) $\sigma_{\bar{x}}^2 = (10-15)^2 \cdot \dfrac{10}{28} + (11-15)^2 \cdot \dfrac{5}{28} + (14-15)^2 \cdot \dfrac{5}{28}$

$+ (15-15)^2 \cdot \dfrac{1}{28} + (25-15)^2 \cdot \dfrac{5}{28} + (26-15)^2 \cdot \dfrac{1}{28} + (29-15)^2 \cdot \dfrac{1}{28} \approx 41.143$

$\sigma_{\bar{x}} \approx 6.414$.

10.33 (a,b) The 20 possible samples, along with their means, are these:

Sample			Mean	Sample			Mean
6	9	12	9	9	12	15	12
6	9	15	10	9	12	18	13
6	9	18	11	9	12	21	14
6	9	21	12	9	15	18	14
6	12	15	11	9	15	21	15
6	12	18	12	9	18	21	16
6	12	21	13	12	15	18	15
6	15	18	13	12	15	21	16
6	15	21	14	12	18	21	17
6	18	21	15	15	18	21	18

(c)

x	9	10	11	12	13	14	15	16	17	18
$f(x)$	$\frac{1}{20}$	$\frac{1}{20}$	$\frac{2}{20}$	$\frac{3}{20}$	$\frac{3}{20}$	$\frac{3}{20}$	$\frac{3}{20}$	$\frac{2}{20}$	$\frac{1}{20}$	$\frac{1}{20}$

(d) The probability that the sample mean differs from 13.5 by less than 3 is $\frac{16}{20} = 0.80$.

10.35 The medians of the samples, in the order listed in the previous problem, are 9, 9, 9, 9, 12, 12, 12, 15, 15, 18, 12, 12, 12, 15, 15, 18, 15, 15, 18, 18. The probability distribution for the median is then

x	9	10	11	12	13	14	15	16	17	18
$f(x)$	$\frac{4}{20}$	$\frac{0}{20}$	$\frac{0}{20}$	$\frac{6}{20}$	$\frac{0}{20}$	$\frac{0}{20}$	$\frac{6}{20}$	$\frac{0}{20}$	$\frac{0}{20}$	$\frac{4}{20}$

The probability that the sample median differs from 13.5 by less than 3 is $\frac{12}{20} = 0.60$. The median is much less likely to be close to the target of 13.5.

10.37 (a) It is divided by $\sqrt{\frac{240}{60}} = 2$. (b) It is divided by $\sqrt{\frac{450}{200}} = 1.5$.

(c) It is divided by $\sqrt{\frac{225}{25}} = 3$. (d) It is multiplied by $\sqrt{\frac{640}{40}} = 4$.

10.39 (a) 0.9771; (b) 0.9909; (c) 0.9977;
(d) 0.9511; (e) 0.9499; (f) 0.9488.

10.41 $\sigma_{\bar{x}} = 32.08$.

10.43 (a) $\mu_{\bar{x}} = (125+150+160+160+170+195)\cdot\frac{1}{6} = 160$ and

$\sigma_{\bar{x}}^2 = (125-160)^2\cdot\frac{1}{6}+(150-160)^2\cdot\frac{1}{6}+\cdots+(195-160)^2\cdot\frac{1}{6} \approx 441.67$ so that

$\sigma_{\bar{x}} = \sqrt{441.67} \approx 21.0$.

(b) $\mu_{\bar{x}} = (150+160+160+170)\cdot\frac{1}{4} = 160$ and

$\sigma_{\bar{x}}^2 = (150-160)^2\cdot\frac{1}{4}+(160-160)^2\cdot\frac{1}{4}+\cdots+(170-160)^2\cdot\frac{1}{4} = 50$ so that

$\sigma_{\bar{x}} = \sqrt{50} \approx 7.1$.

10.45 $\mu_{\bar{x}} = (125+195)\cdot\frac{1}{2} = 160$ and $\sigma_{\bar{x}}^2 = (125-160)^2\cdot\frac{1}{2}+(195-160)^2\cdot\frac{1}{2} = 1{,}225$, so that $\sigma_{\bar{x}} = 35$. This standard deviation is greater than the other two.

10.47 $\sigma_{\bar{x}} = \frac{2.4}{5} = 0.48$.

(a) Since $k = \frac{1.2}{0.48} = 2.5$, the probability is at least $1 - \frac{1}{2.5^2} = 0.84$.

(b) Since $z = 2.50$, the probability is $2(0.4938) = 0.9876$.

10.49 $\sigma_{\bar{x}} = \frac{0.025}{4} = 0.00625$. Since $z = \frac{0.01}{0.00625} = 1.60$, the probability is $2(0.4452) = 0.8904$.

10.51 Since $\bar{x} = \frac{6{,}012}{36} \approx 167$ and $z = \frac{167-163}{\frac{18}{6}} \approx 1.33$, the probability is $0.5000 - 0.4082 = 0.0918$.

10.53 Suppose that you try 50 jelly beans. This would require an average weight in the sample in excess of 4.0 grams. The mean weight of a sample of 50 has a standard deviation of $\frac{0.40}{\sqrt{50}} \approx 0.0566$ gram, and the probability that this mean weight exceeds 4.0 gram is $P[\bar{X} > 4.0] = P\left[Z > \frac{(4.0-4.5)}{0.0566}\right] \approx P[Z > -8.44]$, which is greatly in excess of 0.90. Thus, a sample of 50 is more than you need. Since $\frac{200}{4.5}$ is about 45, you know that a sample of 45 will give you about a 50% chance of reaching the total weight of 200 grams. Your desired value of n is somewhere between 45 and 50. You can now repeat this calculation for a number of sample sizes, as follows:

n	Target Mean	Standard Deviation of $\bar{X}$	z	$P[Z>z]$
45	4.44	0.0596	-1.01	0.8438
46	4.35	0.0590	-2.54	0.9945
47	4.26	0.0583	-4.12	0.9999+

The smallest sample size that will work here is $n = 46$.

10.55 $\frac{\sigma}{\sqrt{256}} = \frac{\sigma}{16}$ and $1.25\cdot\frac{\sigma}{\sqrt{400}} = 1.25\cdot\frac{\sigma}{20} = \frac{\sigma}{16}$.

10.57   The mean 15.72 is compared to $\mu = 16$. The standard deviation 1.9958 is compared to $\sigma_{\bar{x}} = \dfrac{4}{\sqrt{5}} \approx 1.79$.

10.59   The ranges are 9, 10, 10, 2, 4, 2, 9, 9, 8, 6, 11, 11, 5, 2, 11, 7, 12, 12, 10, 6, 9, 12, 10, 13, 6, 9, 9, 10, 9, 10, 13, 8, 2, 9, 7, 6, 9, 13, 9, and 6. The odds are 31 to 9 against it.

10.63   (a) $\binom{40}{2}\binom{20}{2}\binom{10}{2}\binom{10}{2} = 780 \cdot 190 \cdot 45 \cdot 45 = 300{,}105{,}000;$

   (b) $\binom{40}{4}\binom{20}{2}\binom{10}{1}\binom{10}{1} = 91{,}390 \cdot 190 \cdot 10 \cdot 10 = 1{,}736{,}410{,}000.$

10.65 $\dfrac{1}{\binom{80}{3}} = \dfrac{6}{80 \cdot 79 \cdot 78} = \dfrac{1}{82{,}160}.$

10.67 $n_1 = \dfrac{60 \cdot 320 \cdot 12}{320 \cdot 12 + 200 \cdot 10 + 400 \cdot 18 + 80 \cdot 20} = 15.7 \text{ or } 16,$

$n_2 = \dfrac{60 \cdot 200 \cdot 10}{14{,}640} = 8,$

$n_3 = \dfrac{60 \cdot 400 \cdot 18}{14{,}640} = 30,$

$n_4 = \dfrac{60 \cdot 80 \cdot 20}{14{,}640} = 7.$

10.69   Since $N = 6$, $N_1 = 3$, $N_2 = 3$, $k = 2$, $n = 2$, $\sigma_1^2 = \dfrac{200}{3}$, and $\sigma_2^2 = \dfrac{200}{3}$, we get $\sigma_{\bar{x}}^2 = \dfrac{(6-2) \cdot 3^2}{2 \cdot 6^2 (3-1)} \cdot \dfrac{200}{3} + \dfrac{(6-2) \cdot 3^2}{2 \cdot 6^2 (3-1)} \cdot \dfrac{200}{3} = \dfrac{100}{3}$, so that $\sigma_{\bar{x}} = \sqrt{\dfrac{100}{3}} \approx 5.8.$

10.71   (a) $\sqrt{\dfrac{150-40}{150-1}} \approx 0.859;$ (b) $\sqrt{\dfrac{80-25}{80-1}} \approx 0.834.$

10.73   Observe that the standard deviation of $\bar{X}$, the average dollar value in the sample, is $\dfrac{4.4}{\sqrt{36}} \approx 0.7333.$

   (a) $P[\bar{X} < 6.5] = P\left[Z < \dfrac{6.5 - 7.5}{0.7333}\right] \approx P[Z < -1.36] = 0.0869.$

   (b) $P[\bar{X} > 9] = P\left[Z > \dfrac{9 - 7.5}{0.7333}\right] \approx P[Z > 2.05] = 0.0202.$

10.75 (a) $\binom{18}{4} = 3,060;$ (b) $\binom{30}{4} = 27,405;$ (c) $\binom{100}{4} = 3,921,225.$

10.77 $\sigma_{\bar{x}} = \dfrac{3.6}{9} = 0.4.$

(a) Since $z = \dfrac{1.0}{0.4} = 2.5$, the probability is 2(0.4938) = 0.9876.

(b) Since $z = \dfrac{0.5}{0.4} = 1.25$, the probability is 2(0.3944) = 0.78888.

CHAPTER 11

Inferences About Means

11.1 (a) With 95% confidence, he can assert that the maximum error is at most
$1.96 \times \dfrac{13.2}{\sqrt{40}} \approx 4.09$ seconds;

(b) 78.4 ± 4.09 or (74.31 seconds, 82.49 seconds).

11.3 (a) With 99% confidence, he can assert that the maximum error is at most
$2.575 \times \dfrac{1.40}{\sqrt{80}} \approx 0.403$ cm;

(b) 3.26 ± 0.403 or (2.857 cm, 3.663 cm).

11.5 With 95% confidence, the maximum error is $E = 1.96 \cdot \dfrac{3.29}{\sqrt{120}} \approx 0.59$.

11.7 (a) $24.15 \pm \dfrac{1.96 \times 3.29}{\sqrt{120}}$ or 24.15 ± 0.589;

(b) $24.15 \pm \dfrac{2.33 \times 3.29}{\sqrt{120}}$ or 24.15 ± 0.700;

(c) $24.15 \pm \dfrac{2.575 \times 3.29}{\sqrt{120}}$ or 24.15 ± 0.773;

(d) $24.15 \pm \dfrac{2.81 \times 3.29}{\sqrt{120}}$ or 24.15 ± 0.844.

11.9 With 0.95 probability, the error will be at most $E = 1.96 \cdot \dfrac{9.4}{\sqrt{150}} \approx 1.50$. The sample has not yet been taken.

11.11 $255.3 \pm \dfrac{1.96 \cdot 48.2}{\sqrt{60}}$, or 255.3 ± 12.20. This can be expressed as 243.1 to 267.5 drinks.

11.13 With 95% confidence, the maximum error is $E = 1.96 \cdot \dfrac{87}{\sqrt{50}} \cdot \sqrt{\dfrac{370}{419}} \approx 22.7$.

11.15 $240.8 \pm 1.96 \cdot \dfrac{10.2}{\sqrt{40}} \cdot \sqrt{\dfrac{160}{199}}$, or 240.8 ± 2.84; $237.96 < \mu < 243.64$ pounds.

11.17 $252.80 \pm 1.645 \times \dfrac{28.90}{\sqrt{250}} \times \sqrt{\dfrac{3300}{3549}}$, or 252.80 ± 2.90, which is $249.90 < \mu < 255.70$. The interval is a little bit smaller.

11.19 The maximum error, using $\sigma = 2.0$ and 95% confidence, will be at most $\dfrac{1.96 \cdot 2.0}{\sqrt{10}} \approx 1.24$ ounces.

11.21 $n = \left[\dfrac{1.645 \cdot 8.0}{2.2}\right]^2 \approx 36$.

11.23 Use $n \geq \left(\dfrac{1.96 \cdot 8}{2.5}\right)^2 \approx 39.34$. He should increase n to the next integer, requiring a sample of size 40.

11.25 $n = \left(\dfrac{1.645\sigma}{0.3\sigma}\right)^2 \approx 31$.

11.27 (a) $58.22 \pm 2.110 \cdot \dfrac{4.80}{\sqrt{18}}$ or 58.22 ± 2.39 gals. The value 2.110 is $t_{0.025}$ for 17 degrees of freedom. You can give the interval as (55.83, 60.61 gals.).
   (b) $58.22 \pm 2.898 \cdot \dfrac{4.80}{\sqrt{18}}$, or 58.22 ± 3.28. The value 2.898 is $t_{0.005}$ for 17 degrees of freedom. You can give the interval as (54.94, 61.50).
   (c) 2.39 gallons.
   (d) 3.28 gallons.

11.29 The length of the 95% interval is $2t_{0.025} \dfrac{s}{\sqrt{n}}$ and the length of the 99% interval is $2t_{0.005} \dfrac{s}{\sqrt{n}}$. The excess length, expressed as a fraction is just $\dfrac{t_{0.005} - t_{0.025}}{t_{0.025}}$, which is $\dfrac{t_{0.005}}{t_{0.025}} - 1$.
   (a) Using 9 degrees of freedom, the excess length is $\dfrac{3.250}{2.262} - 1 \approx 44\%$.
   (b) Using 19 degrees of freedom, the excess length is $\dfrac{2.861}{2.093} - 1 \approx 37\%$.
   (c) Using 29 degrees of freedom, the excess length is $\dfrac{2.756}{2.045} - 1 \approx 35\%$.

11.31 (a) With 95% confidence, one can say that the maximum error at most is $\dfrac{2.201 \cdot 0.52}{\sqrt{12}} \approx 0.33$ micrograms/cubic foot;
   (b) 2.58 ± 0.33 micrograms/cubic foot.

11.33 (a) $27.33 \pm 2.201 \cdot \dfrac{4.28}{\sqrt{12}}$, or 27.33 ± 2.72; $24.61 < \mu < 30.05$ increased beats per minute.
   (b) $27.33 \pm 3.106 \cdot \dfrac{4.28}{\sqrt{12}}$, or 27.33 ± 3.84; $23.49 < \mu < 31.17$ increased beats per minute.

11.35 Since $n = 4$, $\bar{x} = 14.30$, and $s = 0.0365$, we can assert with 99% confidence that the maximum error is $5.841 \cdot \dfrac{0.0365}{\sqrt{4}} \approx 0.11$ gram.

11.37 Since $\mu_0 = 3{,}600$ and $\sigma_o = 130$, we get $z = \dfrac{3{,}500 - 3{,}600}{130} \approx -0.77$ and a probability of 0.2794.

11.39 The posterior mean is $360.02, and the posterior variance is $\dfrac{1}{\frac{1}{\sigma_0^2}+\frac{n}{\sigma^2}} = \dfrac{1}{\frac{1}{12^2}+\frac{10}{120^2}} \approx 130.9243$, so that the posterior standard deviation is $\sqrt{130.9243} \approx 11.44$. The probability in this distribution between $360 and $400 is the probability in a standard normal distribution between $z = \dfrac{360-360.02}{11.44} \approx 0.00$ and $z = \dfrac{400-360.02}{11.44} \approx 3.49$. This probability is 0.4990, to four figures.

11.41 (a) $\mu_1 = \dfrac{\frac{40}{7.4^2} \cdot 72.9 + \frac{1}{1.5^2} \cdot 65.2}{\frac{40}{7.4^2}+\frac{1}{1.5^2}} \approx 70.0$.

(b) $\dfrac{1}{\sigma_1^2} = \dfrac{40}{7.4^2}+\dfrac{1}{1.5^2} \approx 1.175$, so that $\sigma_1 = 0.92$. Since $z = \dfrac{63-70}{0.92} \approx -7.61$ and $z = \dfrac{68-70}{0.92} \approx -2.17$, the probability is $0.5000 - 0.4850 = 0.0150$.

11.43 (a) Since $z = \dfrac{560-800}{160} = -1.5$ and $z = \dfrac{1,040-800}{160} = 1.5$, the probability is $0.4332 + 0.4332 = 0.8664$.

(b) $\mu_1 = \dfrac{\frac{28}{32^2} \cdot 710 + \frac{1}{160^2} \cdot 800}{\frac{28}{32^2}+\frac{1}{160^2}} \approx \dfrac{19.445}{0.02738} \approx \$710.19$.

(c) $\dfrac{1}{\sigma_1^2} = \dfrac{28}{32^2}+\dfrac{1}{160^2} \approx 0.02738$ so that $\sigma_1 \approx 6.04$. Since $z = \dfrac{560-710.19}{6.04} \approx -246$ $z = \dfrac{1,040-710.19}{6.04} \approx 54.6$, the probability is 100%.

11.45 (a) The police department should use the alternative hypothesis $\mu_2 > \mu_1$ and switch to the radial tires only if the null hypothesis can be rejected.
(b) The police department should use the alternative hypothesis $\mu_2 < \mu_1$ and switch to the radial tires unless the null hypothesis can be rejected.
(c) The police department should use the alternative hypothesis $\mu_2 \neq \mu_1$.

11.47 (a) reject the null hypothesis (error);
(b) reject the null hypothesis (correct);
(c) reject the null hypothesis (error);
(d) reject the null hypothesis (correct).

11.49 If it erroneously rejects the null hypothesis, the testing service commits a Type I error. If it erroneously accepts the null hypothesis, the testing service commits a Type II error.

11.51 Use the null hypothesis that the anti-pollution device for cars is not effective.

11.53 (a) Since $z = \dfrac{6,200-6,000}{\frac{800}{\sqrt{150}}} \approx 3.06$, the probability of Type I error is $0.5000 - 0.4989 = 0.0011$.

(b) Since $z = \dfrac{6,200-6,050}{\frac{800}{\sqrt{150}}} \approx 2.30$, the probability of Type II error is $0.5000 + 0.4893 = 0.9893$.

11.55 (a) The standard deviation of $\bar{x}$ is $\dfrac{8}{\sqrt{100}} = 0.8$. Using $\mu = 65$, the probability of a Type I error is found as $P[\bar{x} \geq 66] = P\left[z \geq \dfrac{66-65}{0.8}\right] = P[z \geq 1.25] = 0.1056$.

(b) Using $\mu = 66.50$, the probability of a Type II error is
$P[\bar{x} < 66] = P\left[z < \dfrac{66-65.50}{0.8}\right] = P[z < 0.625] = 0.7340$.

(c) Using $\mu = 66.50$, the probability of a Type II error is
$P[\bar{x} < 66] = P\left[z < \dfrac{66-66.50}{0.8}\right] = P[z < -0.625] = 0.2660$.

11.57 (a) Since $z = \dfrac{41.0-42.5}{1.07} \approx -1.40$ and $z = \dfrac{44.0-42.5}{1.07} \approx 1.40$, the probability of a Type I error is $2(0.5000-0.4192) = 0.1616$.

(b) For $\mu = 38.5$, $z = \dfrac{41.0-38.5}{1.07} \approx 2.34$ and $z = \dfrac{44.0-38.5}{1.07} \approx 5.14$, and the probability of a Type II error is $0.5000-0.4904 = 0.0096$; by symmetry, the probability is the same for $\mu = 46.5$.

For $\mu = 39.5$, $z = \dfrac{41.0-39.5}{1.07} \approx 1.40$ and $z = \dfrac{44.0-39.5}{1.07} \approx 4.20$, and the probability of a Type II error is $0.5000-0.4192 = 0.0808$; by symmetry, the probability is the same for $\mu = 45.5$.

For $\mu = 40.5$, $z = \dfrac{41.0-40.5}{1.07} \approx 0.47$ and $z = \dfrac{44.0-40.5}{1.07} \approx 3.27$, and the probability of a Type II error is $0.5000-0.1808 = 0.3192$; by symmetry, the probability is the same for $\mu = 44.5$.

For $\mu = 41.5$, $z = \dfrac{41.0-41.5}{1.07} \approx -0.47$ and $z = \dfrac{44.0-41.5}{1.07} \approx 2.34$, and the probability of a Type II error is $0.1808+0.4904 = 0.6712$; by symmetry, the probability is the same for $\mu = 43.5$.

11.59 The term "statistically significant" applies to the difference between the sample means and not to the difference between the population means. For the population means, there is either a difference or no difference.

11.61 It does not make sense since we are not dealing with sample data. There is a difference and that is all there is to it.

11.63 Purely by chance, $40(0.05) = 2$ times can be expected. So 7 times is worrisome.

11.65 (a) If μ represents the true mean number of lectures missed per semester, the null hypothesis should be $\mu = 3.4$ and the alternative should be $\mu \neq 3.4$.

(b) Two-tailed test.

11.67 The null hypothesis is $\mu = 90$ minutes, and the alternative hypothesis $\mu > 90$ minutes.

Modern Elementary Statistics – 9th Edition

11.69   Her results may be that $\bar{x}$ is significantly different from 24 ounces, but the actual average $\bar{x} = 24.08$ ounce is so close that we really are unconcerned. This is a case in which statistical significance is obtained with a result that has no practical significance.

11.71 (a)   1. Test $\mu = 8.8$ vs. alternative $\mu \ne 8.8$.
    2. $\alpha = 0.01$.
    3. Reject the null hypothesis if $|z| \ge z_{0.005} = 2.58$ where $z = \dfrac{\bar{x} - 8.8}{\frac{\sigma}{\sqrt{70}}}$.
    4. The numeric value of z is $\dfrac{8.2 - 8.8}{\frac{2.2}{\sqrt{70}}} \approx -2.28$.
    5. Since $|-2.28| < 2.58$, the null hypothesis can not be rejected.

   (b) $P[|z| > 2.28] = 0.0113 \times 2 = 0.0226$; we have $p = 0.0226$.
   (c)   Since $p > \alpha$, the null hypothesis can not be rejected.

11.73 (a)   1. $H_0: \mu = 40$ and $H_A: \mu \ne 40$.
    2. $\alpha = 0.05$.
    3. Reject the null hypothesis if $|z| \ge 1.96$.
    4. $z = \dfrac{42 - 40}{\frac{2.1}{\sqrt{34}}} \approx 5.55$.
    5. The null hypothesis must be rejected.

   (b)   Since $P[|z| \ge 5.55]$ is less than 0.0000001, we simply say $p < 0.0000001$.
   (c)   Since $p < \alpha$, the null hypothesis is rejected.

   It should be noted that we recommend the use of the sample standard deviation s rather than the claimed standard deviation $\sigma = 0.18$ gram. If you interpret the phrase "typical weight" as including the historical standard deviation, then you should use σ; in such a case the null hypothesis should be amended to read "$\mu = 0.85$ and $\sigma = 0.18$".

11.75   Since 0.005 is less than 0.01, the null hypothesis must be rejected.

11.77   1. $H_0: \mu = 2.41$ and $H_A: \mu \ne 2.41$.
   2.   (a) $\alpha = 0.05$;   (b) $\alpha = 0.01$.
   3.   (a) Reject the null hypothesis if $|z| \ge 1.96$.
        (b) Reject the null hypothesis if $|z| \ge 2.575$.
   4. $z = \dfrac{2.44 - 2.41}{\frac{0.07}{\sqrt{30}}} \approx 2.35$.
   5.   (a) The null hypothesis must be rejected.
        (b) The null hypothesis cannot be rejected.
   (c)   Since $P[|z| \ge 2.35] = 0.0188$, we have $p = 0.0188$.
   (d)   Since $p < 0.02$, the null hypothesis would have been rejected at 0.02 level of significance.

11.79 (a)   We need $n = \dfrac{14^2(1.96 + 0.674)^2}{(260 - 240)^2} \approx 3.4$, or 4 rounded up to the next integer.

   (b)   Yes.

11.81
1. $H_0: \mu = 6.0$ and $H_A: \mu > 6.0$.
2. $\alpha = 0.01$.
3. Reject null hypothesis if $t \geq 2.896$.
4. $t = \dfrac{6.2 - 6.0}{\frac{0.15}{\sqrt{9}}} = 4.00$.
5. The null hypothesis must be rejected; the machine is overfilling cups.

11.83
1. $H_0: \mu = 42$ and $H_A: \mu < 42$.
2. $\alpha = 0.05$.
3. Reject the null hypothesis if $t \leq -1.729$.
4. $t = \dfrac{44.5 - 42}{\frac{2.4}{\sqrt{20}}} \approx 4.66$.
5. The null hypothesis cannot be rejected.

11.85
1. $H_0: \mu = 12.0$ and $H_A: \mu > 12.0$.
2. $\alpha = 0.01$.
3. Reject the null hypothesis if $t \geq 3.747$.
4. $t = \dfrac{12.6 - 12.0}{\frac{0.51}{\sqrt{5}}} \approx 2.63$.
5. The null hypothesis cannot be rejected.

11.87
1. Test $\mu = 600$ versus alternative $\mu < 600$.
2. $\alpha = 0.05$.
3. Reject the null hypothesis if $t \leq -2.015$, where $t = \dfrac{\bar{x} - 600}{\frac{s}{\sqrt{6}}}$ and $2.015 = t_{0.05}$ for 5 degrees of freedom.
4. Numerically, find $\bar{x} = 598.7$ and $s = 22.39$; then $t = \dfrac{598.7 - 600}{\frac{22.39}{\sqrt{6}}} \approx -0.14$.
5. Since −0.14 is not less than or equal to −2.015, the null hypothesis cannot be rejected. Observe that this teacher is not able to reject the claim that the mean reading speed is at least 600 words per minute. He used, however, a very weak experiment to make his point.

11.89 (a)
1. Test $\mu = 60$ versus alternative $\mu \neq 60$.
2. $\alpha = 0.05$.
3. Reject the null hypothesis if $|t| \geq 2.776$, where $t = \dfrac{\bar{x} - 60}{\frac{s}{\sqrt{5}}}$ and $2.776 = t_{0.025}$ for 4 degrees of freedom.
4. Numerically, find $\bar{x} = 64$ and $s = 1.58$; then $t = \dfrac{64 - 60}{\frac{1.58}{\sqrt{5}}} \approx 5.67$.
5. Since $|5.67| > 2.776$, the null hypothesis must be rejected.

(b) Changing the fifth value from 65 to 80 makes $\bar{x} = 67$ and $s = 7.42$. These changes make $t = 2.11$, and the null hypothesis cannot be rejected.

(c) Even though the distance between $\bar{x}$ and 60 has increased, so has the value of s.

11.91 (a) The probability that all tests will be non-significant is $0.90^{20} \approx 0.1216$, so the probability that at least one will "prove" effective is $1 - 0.1216 = 0.8784$, about 88%.
(b) The probability of exactly one significant result is given by the binomial formula $\binom{20}{1} 0.10^1 \, 0.90^{19} \approx 0.2702$.
The probability of more than one significant result is $0.8784 - 0.2702 = 0.6082$, about 61%.

11.93 The tail probability is 0.3668; since 0.3668 exceeds 0.03, the null hypothesis $\mu \geq 600$ could not have been rejected.

11.95
1. $H_0: \mu_1 = \mu_2$ and $H_A: \mu_1 \neq \mu_2$.
2. $\alpha = 0.05$.
3. Reject the null hypothesis if $z \leq -1.96$ or $z \geq -1.96$.
4. $z = \dfrac{9.4 - 7.9}{\sqrt{\dfrac{3.3^2}{40} + \dfrac{2.9^2}{50}}} \approx 2.26$.
5. The null hypothesis must be rejected; the difference between the two sample means is significant.

11.97
1. Test $\mu_m \leq \mu_w + 30$ versus alternative $\mu_m > \mu_w + 30$.
2. $\alpha = 0.01$.
3. Reject the null hypothesis if $z \geq 2.33$, where $z = \dfrac{\overline{x}_m - (\overline{x}_w + 30)}{\sqrt{\dfrac{\sigma_m^2}{n_m} + \dfrac{\sigma_w^2}{n_w}}}$.
4. Replacing σ_m by $s_m = 35.20$ and σ_w by $s_w = 32.65$, we find $z = \dfrac{422.18 - (381.66 + 30)}{\sqrt{\dfrac{35.20^2}{75} + \dfrac{32.65^2}{60}}} \approx 0.31$.
5. Since 0.31 is not greater than or equal to 2.33, the null hypothesis cannot be rejected. We conclude that the men's wages do not significantly exceed the women's wages by $30 or more.

11.99
1. $H_0: \mu_1 - \mu_2 \leq 0.050$ and $H_A: \mu_1 - \mu_2 > 0.050$.
2. $\alpha = 0.01$.
3. Reject the null hypothesis if $z \geq 2.33$.
4. $z = \dfrac{0.135 - 0.082 - 0.050}{\sqrt{\dfrac{0.004^2}{35} + \dfrac{0.005^2}{35}}} \approx 2.78$.
5. The null hypothesis must be rejected; reject the claim.

11.101 $353.78 - 315.14 \pm 2.575 \cdot \sqrt{\dfrac{18.03^2}{80} + \dfrac{21.33^2}{80}}$, or 38.64 ± 8.04; $30.60 < \delta < 46.68$.

11.103
1. $H_0: \mu_1 = \mu_2$ and $H_A: \mu_1 \neq \mu_2$.
2. $\alpha = 0.05$.
3. Reject the null hypothesis if $t \leq -2.074$ or $t \geq -2.074$.
4. $t = \dfrac{41.2 - 45.8}{\sqrt{\dfrac{11(5.2)^2 + 11(6.7)^2}{22} \cdot \left(\dfrac{1}{12} + \dfrac{1}{12}\right)}} \approx -1.88$.
5. The null hypothesis cannot be rejected.

11.105 1. $H_0: \mu_1 = \mu_2$ and $H_A: \mu_1 \neq \mu_2$.
2. $\alpha = 0.05$.
3. Reject the null hypothesis if $t \leq -2.447$ or $t \geq 2.447$.
4. $t = \dfrac{514 - 487}{\sqrt{\dfrac{3(32)^2 + 3(27)^2}{6} \cdot \left(\dfrac{1}{4} + \dfrac{1}{4}\right)}} \approx 1.29$.
5. The null hypothesis cannot be rejected; the difference between the two sample means is not significant.

11.107 1. $H_0: \mu_1 - \mu_2 \leq 10$ and $H_A: \mu_1 - \mu_2 > 10$.
2. $\alpha = 0.05$.
3. Reject the null hypothesis if $t \geq 1.734$.
4. $t = \dfrac{(81 - 61) - 10}{\sqrt{\dfrac{9 \cdot (5.3955)^2 + 9 \cdot (3.3665)^2}{18} \cdot \left(\dfrac{1}{10} + \dfrac{1}{10}\right)}} \approx 4.98$.
5. The null hypothesis must be rejected.

11.109 $514 - 487 \pm 2.447 \cdot \sqrt{\dfrac{3(32)^2 + 3(27)^2}{6} \cdot \left(\dfrac{1}{4} + \dfrac{1}{4}\right)}$, or 27 ± 51.2; $-24.2 < \delta < 78.2$ square feet.

11.111 1. $H_0: \mu_1 = \mu_2$ and $H_A: \mu_1 \neq \mu_2$.
2. $\alpha = 0.01$.
3. Reject the null hypothesis if $|t| \geq 3.25$.
4. $t = \dfrac{-0.02}{\dfrac{0.0287}{\sqrt{10}}} \approx -2.204$.
5. The null hypothesis cannot be rejected. We need the assumptions that the populations are normally distributed.

11.113 1. $H_0: \mu = 85$ and $H_A: \mu < 85$.
2. $\alpha = 0.01$.
3. Reject the null hypothesis if $z \leq -2.33$.
4. $z = \dfrac{76.4 - 85}{\dfrac{12.8}{\sqrt{60}}} \approx -5.2$.
5. The null hypothesis must be rejected. The average useful life is reduced by their use in tropical climate.

11.115 $\bar{x} = 20.17$ and $s = 4.02$, so that the confidence limits are $20.17 \pm 2.571 \cdot \dfrac{4.02}{\sqrt{6}}$, or 20.17 ± 4.22; the confidence interval is $15.95 < \mu < 24.39$.

11.119 (a) This is the probability in a standard normal distribution between $z = \dfrac{4{,}500 - 4{,}500}{280} = 0$ and $z = \dfrac{5{,}000 - 4{,}500}{280} \approx 1.79$, and this probability is 0.4633.

(b) The posterior mean is $\mu_1 = \dfrac{\frac{10}{380^2} \cdot 4{,}702 + \frac{1}{280^2} \cdot 4{,}500}{\frac{10}{380^2} + \frac{1}{280^2}} \approx 4{,}677.36$ and the posterior variance is given through $\dfrac{1}{\sigma_1^2} = \dfrac{10}{380^2} + \dfrac{1}{280^2} \approx 0.0000817551$, so that $\sigma_1 \approx 110.60$.

(c) This is the probability in a standard normal distribution between $z = \dfrac{4{,}500 - 4{,}677.36}{110.60} \approx -1.60$ and $z = \dfrac{5{,}000 - 4{,}677.36}{110.60} \approx 2.92$, and this probability is 0.9434.

(d) The two analyses used different prior distributions and hence come up with different answers. The answers may look close, but realize that the probabilities outside the intervals, namely $1 - 0.9736 = 0.0264$ and $1 - 0.9434 = 0.0566$ are approximately in a 1- to-2 ratio.

11.121 1. $H_0: \mu \leq -1.5$ and $H_A: \mu > -1.5$.
        2. $\alpha = 0.05$.
        3. Reject the null hypothesis if $t \leq -1.761$.
        4. $\bar{x} = -2.13$ and $s = 2.72$ and $t = \dfrac{-2.13 - (-1.5)}{\frac{2.72}{\sqrt{15}}} \approx -0.90$.
        5. The null hypothesis cannot be rejected.

11.123 Find $\bar{x} = 99.7$ and $s = 0.381$. We are 95 percent confident that the error will be at most $\dfrac{2.776 \cdot 0.381}{\sqrt{5}} \approx 0.473$ pound. Here 2.776 is the value of $t_{0.025}$ on 4 degrees of freedom.

11.125 No, since we are dealing with actual votes and not sample data.

11.127 (a) 1. $H_0: \mu = 3.00$ and $H_A: \mu < 3.00$.
        2. $\alpha = 0.01$.
        3. Reject the null hypothesis if $z \leq -2.33$.
        4. $z = \dfrac{2.96 - 3.00}{\frac{0.11}{\sqrt{50}}} \approx -2.57$.
        5. The null hypothesis must be rejected. The commission has grounds upon which to proceed against the manufacturer.

(b) The tail probability is $p = 0.5000 - 0.4949 = 0.0051$. Since $0.0051 > 0.002$. The null hypothesis could not have been rejected at the 0.002 level of significance.

11.129 1. $H_0: \mu_1 - \mu_2 = 3$ and $H_A: \mu_1 - \mu_2 \neq 3$.
        2. $\alpha = 0.05$.
        3. Reject the null hypothesis if $z \leq -1.96$ or $z \geq 1.96$.
        4. $z = \dfrac{67.4 - 62.8 - 3}{\sqrt{\frac{5^2}{50} + \frac{4.6^2}{50}}} \approx 1.67$.
        5. The null hypothesis cannot be rejected; there is no real evidence to support the conjecture.

11.131
1. $H_0: \mu \geq 28$ and $H_A: \mu < 28$.
2. $\alpha = 0.05$.
3. Reject the null hypothesis if $t \leq -1.833$.
4. $t = \dfrac{27.1 - 28}{\frac{1.5542}{\sqrt{10}}} \approx -1.831$.
5. The null hypothesis cannot be rejected; because the t-value of -1.831 is greater than -1.833. However, it is extremely close and we may want to defer judgment.

11.133 (a) She should use the null hypothesis $\mu = 48$ (or $\mu \leq 48$) and the alternative hypothesis $\mu > 48$, and she would prove her point if the null hypothesis can be rejected.

(b) She should use the null hypothesis $\mu \geq 48$ and the alternative hypothesis $\mu < 48$, and she would prove her point if the null hypothesis can be accepted.

CHAPTER 12

Inferences About Standard Deviations

12.1 The interval for σ^2 is $\dfrac{(n-1)s^2}{\chi^2_{0.025}} < \sigma^2 < \dfrac{(n-1)s^2}{\chi^2_{0.975}}$, which in this case is

$\dfrac{17 \cdot 0.014^2}{30.191} < \sigma^2 < \dfrac{17 \cdot 0.014^2}{7.564}$, using 17 degrees of freedom. The interval is

$0.00011036 < \sigma^2 < 0.000441$; in the terms of σ the refractive index is $0.0105 < \sigma < 0.0210$.

12.3 $\dfrac{7(19,200)^2}{16.013} < \sigma^2 < \dfrac{7(19,200)^2}{1.690}$; $12,694 < \sigma < 39,076$ car crossings, which rounds to $12,700 < \sigma < 39,100$ car crossings.

12.5 (a) $\dfrac{7 \cdot 0.56^2}{20.278} < \sigma^2 < \dfrac{7 \cdot 0.56^2}{0.989}$, using 7 d.f., which is $0.108255 < \sigma^2 < 2.219616$, or $0.329 < \sigma < 1.490$ micrograms.

(b) $\dfrac{8 \cdot 0.15^2}{21.955} < \sigma^2 < \dfrac{8 \cdot 0.15^2}{1.344}$, using 8 d.f., which is $0.008199 < \sigma^2 < 0.133929$, or

$0.091 < \sigma < 0.366$ ounce.

12.7 (a) $\dfrac{13.2}{1+\frac{1.96}{\sqrt{80}}} < \sigma < \dfrac{13.2}{1-\frac{1.96}{\sqrt{80}}}$.

$10.83 < \sigma < 16.90$ seconds.

(b) $\dfrac{1.40}{1+\frac{1.96}{\sqrt{160}}} < \sigma < \dfrac{1.40}{1-\frac{1.96}{\sqrt{160}}}$.

$1.21 < \sigma < 1.66$ centimeters.

12.9 The interval for σ is $\dfrac{0.3}{1+\frac{1.96}{\sqrt{120}}} < \sigma < \dfrac{0.3}{1-\frac{1.96}{\sqrt{120}}}$ or $0.254 < \sigma < 0.365$. By squaring, the interval for σ^2 is $0.065 < \sigma^2 < 0.133$ ounce.

12.11 The estimate of σ based on the range is $\dfrac{26-21}{2.53} \approx 1.98$ minutes, compared to $s = 1.79$ minutes.

12.13 The estimate of σ based on the range is $\dfrac{66-62}{2.33} \approx 1.72$ ounces, compared to $s = 1.58$ ounces.

12.15 1. $H_0: \sigma = 2.0$ and $H_A: \sigma < 2.0$.

2. $\alpha = 0.05$.

3. Reject the null hypothesis if $\chi^2 \leq 10.117$.

4. χ^2 is $\dfrac{19 \cdot (1.9)^2}{(2)^2} \approx 17.15$.

5. The null hypothesis cannot be rejected.

12.17 1. $H_0: \sigma \geq 0.040$ and $H_A: \sigma < 0.40$.

2. $\alpha = 0.05$.

3. Reject the null hypothesis if $\chi^2 \leq 0.711$.

4. $\chi^2 = \dfrac{4(0.274)^2}{(0.40)^2} \approx 1.88$.

5. The null hypothesis cannot be rejected.

12.19 The tail value is 0.0537.

12.21 1. $H_0: \sigma = 2.5$ versus $H_A: \sigma \neq 2.5$.
2. $\alpha = 0.05$.

3. Reject the null hypothesis if $z \leq -1.96$ or $z \geq 1.96$, where $z = \dfrac{s - \sigma}{\frac{\sigma}{\sqrt{2n}}}$.

4. z is $\dfrac{2.1 - 2.5}{\frac{2.5}{\sqrt{68}}} \approx -1.32$.

5. The null hypothesis cannot be rejected.

12.23 1. $H_0: \sigma_1 = \sigma_2$ and $H_A: \sigma_1 < \sigma_2$.

2. $\alpha = 0.05$.

3. Reject the null hypothesis if $F \geq 2.72$.

4. F is $\dfrac{(4.4)^2}{(2.6)^2} \approx 2.86$.

5. The null hypothesis must be rejected; the first technique is less variable than the second.

12.25 1. $H_0: \sigma_G = \sigma_R$ versus $H_A: \sigma_G \neq \sigma_R$.

2. $\alpha = 0.02$.

3. Reject the null hypothesis if $F \geq 6.63$.

4. $F = \dfrac{(15.4920)^2}{(10.4403)^2} \approx 2.202$.

5. The null hypothesis cannot be rejected.

12.27 1. $H_0: \sigma_1 = \sigma_2$ and $H_A: \sigma_1 \neq \sigma_2$.
2. $\alpha = 0.05$.
3. Reject the null hypothesis if $F \geq 2.20$.
4. $F = \dfrac{(9.42)^2}{(5.48)^2} \approx 2.95$.
5. The null hypothesis must be rejected.

12.29 (a) $F_{0.95} = \dfrac{1}{3.29} \approx 0.304$;   (b) $F_{0.99} = \dfrac{1}{5.61} \approx 0.178$.

12.31 1. $H_0: \sigma_1 = \sigma_2$ and $H_A: \sigma_1 \neq \sigma_2$.
2. $\alpha = 0.10$.
3. Reject the null hypothesis if $F \geq 3.68$.
4. $F = \dfrac{(4.6961)^2}{(2.4681)^2} \approx 3.62$.
5. The null hypothesis cannot be rejected.

12.33 (a) 3.05 minutes; (b) $\dfrac{30.0 - 25.15}{1.35} \approx 3.59$ minutes; (c) $\dfrac{33.0 - 22.4}{3.26} \approx 3.25$ minutes.

12.35 1. $H_0: \sigma_1 = \sigma_2$ and $H_A: \sigma_1 < \sigma_2$.
2. $\alpha_2 = 0.05$.
3. Reject the null hypothesis if $F \geq 2.82$.
4. $F = \dfrac{(6.7)^2}{(5.2)^2} \approx 1.66$.
5. The null hypothesis cannot be rejected.

12.37 1. $H_0: \sigma = 0.3$ and $H_A: \sigma > 0.3$.
2. $\alpha = 0.05$.
3. Reject the null hypothesis if $\chi^2 \geq 24.996$, where 24.996 is $\chi^2_{0.05}$ using 15 degrees of freedom, and χ^2 is $\dfrac{15 \cdot s^2}{0.3^2}$.
4. The numeric value of χ^2 is $\dfrac{15 \cdot 0.38^2}{0.3^2} \approx 24.07$.
5. Since 24.07 is less than 24.996, the null hypothesis cannot be rejected.

CHAPTER 13

Inferences About Proportions

13.1 (a) $0.02 < p < 0.98$; (b) $0.00625 < p < 0.99375$;
(c) $n > 125$; (d) $n > 250$.

13.3 (a) The sample proportion opposed to the increase is 0.72, and the 95 percent confidence interval is $0.72 \pm 1.96\sqrt{\frac{0.72 \cdot 0.28}{200}}$, or 0.72 ± 0.062, which may be written as simply (0.658, 0.782).

(b) The sample proportion in favor of the increase is 0.28, and the 95 percent confidence interval is $0.28 \pm 1.96\sqrt{\frac{0.28 \cdot 0.72}{200}}$, or 0.28 ± 0.062, meaning (0.218, 0.342).

(c) The endpoints in (b) can be found directly from the endpoints in (a). We have $0.218 = 1 - 0.782$ and $0.342 = 1 - 0.658$. This is hardly surprising, since problems (a) and (b) are logically equivalent.

13.5 (a) The sample proportion is $\hat{p} = 0.1333$, and the 99 percent confidence interval is $0.1333 \pm 2.58\sqrt{\frac{0.1333 \cdot 0.8667}{180}}$, or 0.1333 ± 0.0654, which may be expressed as simply (0.0679, 0.1987). To three figures, this interval is (0.068, 0.199).

(b) We can be 99% confident that the maximum error is at most 0.0654.

13.7 With 90% confidence, the maximum error is $1.645 \cdot \sqrt{\frac{(0.45)(0.55)}{120}} \approx 0.075$.

13.9 With 99% confidence, the maximum error is $2.575 \cdot \sqrt{\frac{(0.38)(0.62)}{400}} \approx 0.0625$.

13.11 (a) Since $\frac{x}{n} = \frac{176}{312} = 0.564$, we get $0.564 \pm 1.96\sqrt{\frac{0.564 \times 0.436}{312}}$, or 0.564 ± 0.055, which is $0.509 < p < 0.619$.

(b) With 90% confidence, the maximum error is $1.645 \cdot \sqrt{\frac{0.564 \cdot 0.436}{312}} \approx 0.046$.

13.13 (a) $2 \times \frac{1}{2n} = \frac{1}{n}$;

(b) $\hat{p} = \frac{127}{305} \approx 0.416$; $0.416 \pm \left(1.96\sqrt{\frac{0.416 \cdot 0.584}{305}} + \frac{1}{2 \cdot 305}\right)$, or 0.416 ± 0.057 and $0.359 < p < 0.473$.

(c) $\hat{p} = \frac{176}{312} \approx 0.564$; $0.564 \pm \left(1.96\sqrt{\frac{0.564 \cdot 0.436}{312}} + \frac{1}{2 \cdot 312}\right)$, or 0.564 ± 0.0567 and $0.507 < p < 0.621$.

13.15 With 95% confidence, the maximum error is $1.96 \cdot \sqrt{\frac{(0.45)(0.55)}{80}} \approx 0.109$.

13.17 Since $\frac{x}{n} = \frac{34}{100} = 0.34$, we get $0.34 \pm 1.96 \cdot \sqrt{\frac{(0.34)(0.66)(360-100)}{100(360-1)}}$, or 0.34 ± 0.079 and $0.261 < p < 0.419$.

13.19 (a) $n \geq \frac{1.96^2}{4 \cdot 0.02^2} = 2,401$.

(b) $n \geq \frac{1.96^2 \cdot 0.35 \cdot 0.65}{0.02^2} = 2,184.91$, which should be rounded up to 2,185.

13.21 (a) 156; (b) 625;
(c) 2,500; (d) 10,000. This assumes 95% confidence and the approximation used in problem 13.20.

13.23 (a) $\frac{2.33^2}{4 \cdot 0.025^2} = 2,171.56$, which should be rounded up to 2,172.

(b) $\frac{2.33^2 \cdot 0.30 \cdot 0.70}{0.025^2} = 1,824.11$, which should be rounded up to 1,825.

13.25 $\frac{(0.50)(0.206)}{(0.50)(0.206)+(0.50)(0.071)} \approx 0.744$ and $1 - 0.744 = 0.256$.

13.27 (a) $\frac{(0.11)(0.2)}{(0.11)(0.2)+(0.02)(0.4)+(0.01)(0.6)+(0.02)(0.8)+(0.10)1} = \frac{0.022}{0.152} \approx 0.145$;

$\frac{(0.02)(0.4)}{0.152} \approx 0.053$, $\frac{(0.01)(0.6)}{0.152} \approx 0.039$, $\frac{(0.02)(0.8)}{0.152} \approx 0.105$, and $\frac{(0.10)1}{0.152} \approx 0.658$.

(b) $\mu_1 = 0(0.145) + 1(0.053) + 2(0.039) + 3(0.105) + 4(0.658) \approx 3.08$.

13.29 The lower endpoint of the confidence interval is $\hat{p} - z_{\alpha/2}\sqrt{\frac{\hat{p}(1-\hat{p})}{n}}$. If $x = 0$, then this lower endpoint is zero. If $x = 1$, then $p = \frac{1}{n}$, and the lower endpoint is $\frac{1}{n} - z_{\alpha/2}\frac{1}{n}\sqrt{\frac{n-1}{n}} = \frac{1}{n}\left(1 - z_{\alpha/2}\sqrt{\frac{n-1}{n}}\right)$. For $z_{\alpha/2} = 1.96$ (say), the lower endpoint is negative as long as $n \geq 3$.

One can make a similar argument for $x = 2$. This example shows that indeed the left endpoint of the confidence interval can be less than zero.

13.31 1. $H_0: p = 0.10$ and $H_A: p > 0.10$.
2. $\alpha = 0.05$.
3. Reject the null hypothesis if the probability of 5 or more successes is less than or equal to 0.05.
4. The probability of 5 or more successes is $0.022 + 0.005 + 0.001 = 0.028$.
5. The null hypothesis must be rejected.

13.33
1. $H_0: p = 0.20$ versus $H_A: p > 0.20$.
2. $\alpha = 0.05$.
3'. The statistic is x itself; $x = 5$.
4'. The probability of an outcome of 5 or more from the binomial distribution with $n = 12$ and $p = 0.20$ is 0.073, obtained from Table V. This is the p-value.
5'. Since $0.073 > 0.05$, the null hypothesis cannot be rejected.

13.35
1. $H_0: p = 0.50$ and $H_A: p \neq 0.50$.
2. $\alpha = 0.05$.
3. Reject the null hypothesis if the probability of 12 or more successes, or that of 12 or fewer successes, is less than or equal to 0.025.
4. The probability of 12 or more successes is 0.007.
5. The null hypothesis must be rejected.

13.37
1. $H_0: p \geq 0.70$ and $H_A: p < 0.70$.
2. $\alpha \leq 0.05$.
3. Reject the null hypothesis if the probability of 5 or fewer successes is less than or equal to 0.05.
4. The probability of 5 or fewer successes is $0.001 + 0.008 + 0.029 = 0.038$.
5. The null hypothesis must be rejected; the data refute the claim.

13.39 (a) $x \leq 4$ or $x \geq 13$; (b) $x \leq 3$ or $x \geq 13$; (c) $x \leq 2$ or $x \geq 15$.

13.41
1. $H_0: p = 0.30$ and $H_A: p \neq 0.30$.
2. $\alpha = 0.05$.
3. Reject the null hypothesis if $z \leq -1.96$ or $z \geq 1.96$.
4. $z = \dfrac{41 - 150(0.30)}{\sqrt{150(0.30)(0.70)}} \approx -0.71$.
5. The null hypothesis cannot be rejected.

13.43
1. $H_0: p \leq 0.06$ and $H_A: p > 0.06$.
2. $\alpha \leq 0.01$.
3. Reject the null hypothesis if $z \geq 2.33$.
4. $z = \dfrac{17 - 200(0.06)}{\sqrt{200(0.06)(0.94)}} \approx 1.49$.
5. The null hypothesis cannot be rejected.

13.45 (a)
1. $H_0: p = 0.90$ versus $H_A: p \neq 0.90$.
2. $\alpha \leq 0.05$.
3'. The statistic is x itself; $x = 11$.
4'. The probability of an outcome of 11 or less from the binomial distribution with $n = 15$ and $p = 0.90$ is 0.055, and the probability of an outcome of 11, or more is 0.988, obtained from Table V. The p-value is twice the smaller of these, namely $2 \cdot 0.055 = 0.11$.
5'. Since 0.11 exceeds 0.05, the null hypothesis cannot be rejected.

(b) 1. $H_0: p = 0.90$ versus $H_A: p \neq 0.90$.
2. $\alpha \leq 0.05$.
3'. The statistic is x itself; $x = 15$.
4'. The probability of an outcome of 15 or less from the binomial distribution with $n = 15$ and $p = 0.90$ is 1.000, and the probability of an outcome of 15, or more is 0.206, obtained from Table V. The p-value is twice the smaller of these, namely $2 \cdot 0.026 = 0.412$.
5'. Since 0.412 exceeds 0.05, the null hypothesis cannot be rejected.
This experiment simply does not have the ability to detect a p bigger than 0.90. Indeed, a perfect 100 percent success rate in the data is not sufficient to conclude that the success rate is not equal to 0.90.

13.47
1. $H_0: p = 0.5$ and $H_A: p \neq 0.5$.
2. $\alpha = 0.05$.
3. Reject the null hypothesis if $z \leq -1.96$ or $z \geq 1.96$.
4. $z = \dfrac{166 - 400 \times 0.5}{\sqrt{400 \times 0.5 \times 0.5}} \approx -3.4$.
5. The null hypothesis must be rejected.

13.49
1. $H_0: p_1 = p_2$ and $H_A: p_1 < p_2$.
2. $\alpha = 0.01$.
3. Reject the null hypothesis if $z \leq -2.33$.
4. $\hat{p} = \dfrac{412 + 311}{5{,}000 + 3{,}000} = 0.0904$, so that $z = \dfrac{\frac{412}{5{,}000} - \frac{311}{3{,}000}}{\sqrt{(0.0904)(0.9096)\left(\frac{1}{5{,}000} + \frac{1}{3{,}000}\right)}} \approx -3.21$.
5. The null hypothesis must be rejected; the more expensive mail solicitation is more effective.

13.51
1. $H_0: p_1 = p_2$ and $H_A: p_1 \neq p_2$.
2. $\alpha = 0.05$.
3. Reject the null hypothesis if $z \leq -1.96$ or $z \geq 1.96$.
4. $\hat{p} = \dfrac{74 + 86}{200 + 200} = 0.40$, so that $z = \dfrac{\frac{74}{200} - \frac{86}{200}}{\sqrt{(0.40)(0.60)\left(\frac{1}{200} + \frac{1}{200}\right)}} \approx -1.22$.
5. The null hypothesis cannot be rejected; the difference between the two sample proportions is not significant.

13.53
1. $H_0: p_1 = p_2$ versus $H_A: p_1 \neq p_2$.
2. $\alpha = 0.05$.
3. Reject the null hypothesis if $|z| \geq 1.96$.
4. $\hat{p} = \dfrac{98 + 121}{144 + 144} \approx 0.76$, so that $z = \dfrac{\frac{98}{144} - \frac{121}{144}}{\sqrt{(0.76)(0.24)\left(\frac{1}{144} + \frac{1}{144}\right)}} \approx \dfrac{-0.1597}{0.05} \approx -3.19$.
5. The null hypothesis must be rejected.

13.55 (a)
1. $H_0: p_1 - p_2 = 0.20$ and $H_A: p_1 - p_2 > 0.20$.
2. $\alpha = 0.05$.
3. Reject the null hypothesis if $z \geq 1.645$.
4. Since $\dfrac{205}{250} = 0.82$ and $\dfrac{137}{250} = 0.548$, we get $z = \dfrac{\frac{205}{250} - \frac{137}{250} - 0.20}{\sqrt{\frac{(0.82)(0.18)}{250} + \frac{(0.548)(0.452)}{250}}} \approx 1.81$.
5. The null hypothesis must be rejected; the test item appears to be good.

(b) 1. $H_0: p_1 - p_2 = 0.06$ and $H_A: p_1 - p_2 > 0.06$.
2. $\alpha = 0.05$.
3. Reject the null hypothesis if $z \geq 1.645$.
4. Since $\frac{177}{500} = 0.354$ and $\frac{110}{400} = 0.275$, we get $z = \frac{\frac{177}{500} - \frac{110}{400} - 0.06}{\sqrt{\frac{(0.354)(0.646)}{500} + \frac{(0.275)(0.725)}{400}}} \approx 0.61$.
5. The null hypothesis cannot be rejected.

13.57 $\frac{177}{200} - \frac{100}{400} \pm 2.575 \cdot \sqrt{\frac{(0.354)(0.646)}{500} + \frac{(0.275)(0.725)}{400}}$, or 0.079 ± 0.080;
$-0.001 < p_1 - p_2 < 0.159$.

13.59 Odds ratio $\frac{\frac{p_1}{1-p_1}}{\frac{p_2}{1-p_2}} = 1$ if $p_1 = p_2$.

13.61 (a) 1. $H_0: p_1 = p_2$ versus $H_A: p_1 \neq p_2$.
2. $\alpha = 0.05$.
3. Reject the null hypothesis if $|z| \geq 1.96$.
4. $\hat{p} = \frac{36+53}{80+70} \approx 0.59$ and $z = \frac{0.45 - 0.76}{\sqrt{(0.59)(0.41)\left(\frac{1}{80} + \frac{1}{70}\right)}} \approx \frac{-0.31}{0.08} \approx -3.88$.
5. The null hypothesis must be rejected; there is a difference.
(b) Rate difference is $0.45 - 0.76 = -0.31$; odds ratio of comparing freshman to junior is $\frac{0.45/0.55}{0.76/0.24} \approx 0.258$.

13.63 (a) 1. $H_0: p_1 = p_2$ versus $H_A: p_1 \neq p_2$.
2. $\alpha = 0.05$.
3. Reject the null hypothesis if $|z| \geq 1.96$.
4. $\hat{p} = \frac{70+65}{1,020+3,400} \approx 0.03$ and $z = \frac{\frac{70}{1,020} - \frac{65}{3,400}}{\sqrt{(0.03)(0.97)\left(\frac{1}{1,020} + \frac{1}{3,400}\right)}} \approx \frac{0.0495}{0.006} = 8.25$.
5. The null hypothesis must be rejected.
(b) Rate difference is $= \frac{70}{1,020} - \frac{65}{3,400} \approx 0.0495$.
(c) Rate ratio $= \frac{70/1,020}{65/3,400} \approx 3.59$.
(d) Odds ratio $= \frac{70 \cdot 3,335}{65 \cdot 950} \approx 3.78$.
(e) (c) and (d).

13.65 $\chi^2 = \frac{67^2}{46.4} + \frac{64^2}{63.6} + \frac{25^2}{46.0} + \frac{42^2}{51.8} + \frac{76^2}{70.9} + \frac{56^2}{51.3} + \frac{10^2}{20.8} + \frac{23^2}{28.5} + \frac{37^2}{20.7} - 400 \approx 40.9$.

13.67 The null hypothesis is that the probabilities of the four response categories (go down, remain the same, go up, can't tell) are the same for each of the three types of dealers.

13.69 The null hypothesis is that the probability of favoring the tax increase is the same in each of the four cities.

13.71 1. H_0: the same proportions of freshmen, sophomores, juniors and seniors use the four means of transportation.
H_A: The proportions are not the same for at least one of the four means of transportation.
2. $\alpha = 0.05$.
3. Reject the null hypothesis if $\chi^2 \geq 16.919$.
4. The row totals are 352, 129, 149, and 170; the column totals are all 200; and the grand total is 800. The expected frequencies for the first row are all $\frac{352 \cdot 200}{800} = 88$, those for the second row are all $\frac{129 \cdot 200}{800} = 32.25$, those for the third row are all $\frac{149 \cdot 200}{800} = 37.25$, and those for the fourth row are all $200 - (88 + 32.25 + 37.25) = 42.5$.
$$\chi^2 = \frac{(104-88)^2}{88} + \frac{(87-88)^2}{88} + \cdots + \frac{(39-42.5)^2}{42.5} + \frac{(53-42.5)^2}{42.5} \approx 25.5.$$
5. The null hypothesis must be rejected; the proportions are not the same for at least one of the four means of transportation.

13.73 1. H_0: The size is independent of the number of persons in the shopper's household.
H_A: Tube size is not independent of the number of persons in the shopper's household.
2. $\alpha = 0.01$.
3. Reject the null hypothesis if $\chi^2 \geq 16.812$.
4. The row totals are 260, 106, and 146; the column totals are 108, 209, 133, and 62; and the grand total is 512. The expected frequencies for the first row are $\frac{260 \cdot 108}{512} \approx 54.8$, $\frac{260 \cdot 209}{512} = 106.1$, $\frac{206 \cdot 133}{512} \approx 67.5$, and $260 - (54.8 + 106.1 + 67.5) = 31.6$; those for the second row are $\frac{106 \cdot 108}{512} \approx 22.4$, $\frac{106 \cdot 209}{512} \approx 43.3$, $\frac{106 \cdot 133}{512} \approx 27.5$, and $106 - (22.4 + 43.3 + 27.5) = 12.8$, and those for the third row are $108 - (54.8 + 22.4) = 30.8$, $209 - (106.1 + 43.3) = 59.6$, $133 - (67.5 + 27.5) = 38.0$, and $62 - (31.6 + 12.8) = 17.6$.
$$\chi^2 = \frac{(23-54.8)^2}{54.8} + \frac{(116-106.1)^2}{106.1} + \cdots + \frac{(8-17.6)^2}{17.6} \approx 88.9.$$
5. The null hypothesis must be rejected; tube size is not independent of the number of persons in the shopper's household.

13.75 1. H_0: Parent's reaction to the course is independent of the number of children they have in school.
H_A: Parents' reaction to the course is not independent of the number of children they have in school.
2. $\alpha = 0.05$.
3. Reject the null hypothesis if $\chi^2 \geq 9.488$.
4. The row totals are 100, 137, and 123; the column totals are 160, 139, and 61; and the grand total is 360. The expected frequencies for the first row are $\frac{100 \cdot 160}{360} \approx 44.4$, $\frac{100 \cdot 139}{360} \approx 38.6$, and $100 - (44.4 + 38.6) = 17.0$; those for the second row are $\frac{137 \cdot 160}{360} \approx 60.9$, $\frac{137 \cdot 139}{360} \approx 52.9$, and $137 - (60.9 + 52.9) = 23.2$; and those for the third row are $160 - (44.4 + 60.9) = 54.7$, $139 - (38.6 + 52.9) = 47.5$, and

$$61 - (17.0 + 23.2) = 20.8.$$
$$\chi^2 = \frac{(48-44.4)^2}{44.4} + \frac{(40-38.6)^2}{38.6} + \cdots + \frac{(46-47.5)^2}{47.5} + \frac{(20-20.8)^2}{20.8} \approx 4.01.$$

5. The null hypothesis cannot be rejected; there is no real evidence of a relationship between family size and the parents' reaction.

13.77
1. H_0: $p_1 = p_2 = p_3$ and H_A: The p's are not all equal.
2. $\alpha = 0.05$.
3. Reject the null hypothesis if $\chi^2 \geq 5.991$.
4. Since $\hat{p} = \frac{44+53+133}{100+100+200} = 0.575$, the expected frequencies for the first row are $100(0.575) = 57.5$, $100(0.575) = 57.5$, and $200(0.575) = 115$, and those for the second row are $100 - 57.5 = 42.5$, $100 - 57.5 = 42.5$, and $200 - 115 = 85$.
$$\chi^2 = \frac{(44-57.5)^2}{57.5} + \frac{(53-57.5)^2}{57.5} + \frac{(133-115)^2}{115} + \frac{(56-42.5)^2}{42.5} + \frac{(47-42.5)^2}{42.5}$$
$$+ \frac{(67-85)^2}{85} \approx 14.92.$$
5. The null hypothesis must be rejected; the differences among the three sample proportions are significant.

13.79
1. H_0: $p_1 = p_2 = p_3$ and H_A: The p's are not all equal.
2. $\alpha = 0.05$.
3. Reject the null hypothesis if $\chi^2 \geq 5.991$.
4. Since $\hat{p} = \frac{8+13+12}{60+80+60} = 0.165$, the expected frequencies for the first row are $60(0.165) = 9.9$, $80(0.165) = 13.2$, and $60(0.165) = 9.9$, and those for the second row are $60 - 9.9 = 50.1$, $80 - 13.2 = 66.8$, and $60 - 9.9 = 50.1$.
$$\chi^2 = \frac{(8-9.9)^2}{9.9} + \frac{(13-13.2)^2}{13.2} + \frac{(12-9.9)^2}{9.9} + \frac{(52-50.1)^2}{50.1} + \frac{(67-66.8)^2}{66.8}$$
$$+ \frac{(48-50.1)^2}{50.1} \approx 0.97.$$
5. The null hypothesis cannot be rejected; the differences among the three sample proportions are not significant.

13.81
1. H_0: $p_1 = p_2 = p_3 = p_4 = p_5$ and H_A: The p's are not all equal.
2. $\alpha = 0.01$.
3. Reject the null hypothesis if $\chi^2 \geq 13.277$.
4. Since $\hat{p} = \frac{74+81+69+75+91}{100+100+100+100+100} = 0.78$, the expected frequencies for the first row are all $100(0.78) = 78$, and those for the second row are all $100 - 78 = 22$.
$$\chi^2 = \frac{(74-78)^2}{78} + \frac{(81-78)^2}{78} + \cdots + \frac{(25-22)^2}{22} + \frac{(8-22)^2}{22} \approx 16.55.$$
5. The null hypothesis must be rejected; the differences among the five sample proportions are significant.

13.85 Not only will each o value be doubled, but also each e value will double. Accordingly, each summoned $\frac{(o-e)^2}{e}$ will also double, and so will χ^2.

13.87 Letting r_i and c_j denote the total of the ith row and the total of the jth column, we get
$$\frac{r_i \cdot c_1}{n} + \frac{r_i \cdot c_2}{n} + \cdots + \frac{r_i \cdot c_c}{n} = \frac{r_i}{n}(c_1 + c_2 + \cdots + c_c) = \frac{r_i}{n} \cdot n = r_i$$ for the sum of the expected frequencies in the ith row.

13.89
1. H_0: The probability is $\frac{1}{6}$ for each face of the die.
   H_A: The probabilities are not all $\frac{1}{6}$.
2. $\alpha = 0.05$.
3. Reject the null hypothesis if $\chi^2 \geq 11.070$.
4. Since the expected frequencies are all equal to $720 \cdot \frac{1}{6} = 120$, we get
$$\chi^2 = \frac{(129-120)^2}{120} + \frac{(107-120)^2}{120} + \frac{(98-120)^2}{120} + \frac{(132-120)^2}{120} + \frac{(136-120)^2}{120}$$
$$+ \frac{(118-120)^2}{120} \approx 9.48.$$
5. The null hypothesis cannot be rejected; there is no real evidence that the die is not balanced and randomly tossed.

13.91
1. H_0: The coins are all balanced and randomly tossed.
   H_A: The coins are not all balanced or are not randomly tossed.
2. $\alpha = 0.05$.
3. Reject the null hypothesis if $\chi^2 \geq 9.488$.
4. Since the expected frequencies are $320 \cdot \frac{1}{16} = 20$, $320 \cdot \frac{4}{16} = 80$, $320 \cdot \frac{6}{16} = 120$, $320 \cdot \frac{4}{16} = 80$, and $320 \cdot \frac{1}{16} = 20$, we get
$$\chi^2 = \frac{(14-20)^2}{20} + \frac{(68-80)^2}{80} + \frac{(108-120)^2}{120} + \frac{(104-80)^2}{80} + \frac{(26-20)^2}{20} = 13.8.$$
5. The null hypothesis must be rejected; the coins are not all balanced or are not randomly tossed.

13.93
1. H_0: Data constitute a sample from a Poisson population with $\lambda = 4.6$.
   H_A: Data do not constitute a sample from a Poisson population with $\lambda = 4.6$.
2. $\alpha = 0.05$.
3. Reject the null hypothesis if $\chi^2 \geq 14.067$.
4. The expected frequencies are $400(0.010) = 4.0$, $400(0.046) = 18.4$, $400(0.107) = 42.8$, $400(0.163) = 65.2$, $400(0.187) = 74.8$, $400(0.173) = 69.2$, $400(0.132) = 52.8$, $400(0.087) = 34.8$, and $400(0.095) = 38.0$. Combining the first two classes, we get
$$\chi^2 = \frac{(14-22.4)^2}{22.4} + \frac{(54-42.8)^2}{42.8} + \frac{(76-65.2)^2}{65.2} + \frac{(68-74.8)^2}{74.8}$$
$$+ \frac{(45-52.8)^2}{52.8} + \frac{(41-34.8)^2}{34.8} + \frac{(26-38.0)^2}{38.0} \approx 15.4.$$
5. The null hypothesis should be rejected; data do not constitute a sample from a Poisson population with $\lambda = 4.6$.

13.95 We begin by calculating the means as $\bar{x} = \dfrac{0 \cdot 105 + 1 \cdot 76 + 2 \cdot 17 + 3 \cdot 2 + 4 \cdot 0}{200} = 0.58$; hence $np = 4p = 0.58$; $p = 0.145$.

1. H_0: Data constitute a sample from binomial population with $n = 4$.
   H_A: Data do not constitute a sample from binomial population with $n = 4$.
2. $\alpha = 0.05$.
3. Reject the null hypothesis if $\chi^2 \geq 3.841$.
4. The expected frequencies are $200(0.5344) = 106.9$, $200(0.3625) = 72.5$, $200(0.092) = 18.4$, $200(0.01) = 2$, and 0.0. Combining the last three classes, we get
$$\chi^2 = \frac{(105 - 106.9)^2}{106.9} + \frac{(76 - 72.5)^2}{72.5} + \frac{(19 - 20.4)^2}{20.4} = 0.3.$$
5. The null hypothesis cannot be rejected.

13.97 (a) Since $z = \dfrac{14.5 - 20.7}{5.4} \approx -1.15$, $z = \dfrac{19.5 - 20.7}{5.4} \approx -0.22$, $z = \dfrac{24.5 - 20.7}{5.4} \approx 0.70$, $z = \dfrac{29.5 - 20.7}{5.4} \approx 1.63$, and $z = \dfrac{34.5 - 20.7}{5.4} \approx 2.56$, the probabilities are $0.5000 - 0.3749 = 0.1251$, $0.3749 - 0.0871 = 0.2878$, $0.0871 + 0.2580 = 0.3451$, $0.4484 - 0.2580 = 0.1904$, $0.4948 - 0.4484 = 0.0464$, and $0.5000 - 0.4948 = 0.0052$.

(b) The expected frequencies are $80(0.1251) \approx 10.0$, $80(0.2878) \approx 23.0$, $80(0.3451) \approx 27.6$, $80(0.1904) \approx 15.2$, $80(0.0464) \approx 3.7$, and $80(0.0052) \approx 0.4$. In what follows, the fourth, fifth, and sixth classes are combined.

(c) 1. H_0: Data constitute a sample from a normal population.
   H_A: Data do not constitute a sample from a normal population.
2. $\alpha = 0.05$.
3. Reject the null hypothesis if $\chi^2 \geq 3.841$.
4. $\chi^2 = \dfrac{(8 - 10.0)^2}{10.0} + \dfrac{(28 - 23.0)^2}{23.0} + \dfrac{(27 - 27.6)^2}{27.6} + \dfrac{(17 - 19.3)^2}{19.3} \approx 1.77$.
5. The null hypothesis cannot be rejected; there is no real evidence that the population is not normal.

13.99 $\bar{x} = \dfrac{0 \cdot 106 + 1 \cdot 98 + 2 \cdot 35 + 3 \cdot 9 + 4 \cdot 2}{250} = 0.812$ hence $np = 4p = 0.812$. $p \approx 0.20$.

1. H_0: Data constitute a sample from binomial population with $n = 4$, $p = 0.20$,
   H_A: Data do not constitute a sample from binomial population with $n = 4$, $p = 0.20$.
2. $\alpha = 0.05$.
3. Reject the null hypothesis if $\chi^2 \geq 5.991$.
4. The expected frequencies are $250(0.41) = 102.5$, $250(0.41) = 102.5$, $250(0.154) = 38.5$, $250(0.026) = 6.5$, $250(0.002) = 0.5$. Combining the last two classes, we get
$$\chi^2 = \frac{(106 - 102.5)^2}{102.5} + \frac{(98 - 102.5)^2}{102.5} + \frac{(35 - 38.5)^2}{38.5} + \frac{(11 - 7)^2}{7} \approx 2.92.$$
5. The null hypothesis cannot be rejected.

13.101 1. $H_0: p_1 = p_2 = p_3$ and H_A: The p's are not all equal.
2. $\alpha = 0.01$.
3. Reject the null hypothesis if $\chi^2 \geq 9.210$.
4. Since $\hat{p} = \dfrac{63 + 84 + 69}{100 + 100 + 100} = 0.72$, the expected frequencies for the first row are all $100(0.72) = 72$, and those for the second row are all $100 - 72 = 28$. Thus,
$$\chi^2 = \dfrac{(63-72)^2}{72} + \dfrac{(84-72)^2}{72} + \dfrac{(69-72)^2}{72} + \dfrac{(37-28)^2}{28} + \dfrac{(16-28)^2}{28} + \dfrac{(31-28)^2}{28} \approx 11.6.$$
5. The null hypothesis must be rejected; the differences among the three sample proportions are significant.

13.103 (a) 1. $H_0: p_1 = p_2$ versus $H_A: p_1 \neq p_2$.
2. $\alpha = 0.01$.
3. Reject the null hypothesis if $|z| \geq 2.575$.
4. Since $\hat{p} = \dfrac{86 + 42}{300 + 200} = 0.256$, we get $z = \dfrac{\tfrac{86}{300} - \tfrac{42}{200}}{\sqrt{(0.256)(0.744)\left(\tfrac{1}{300} + \tfrac{1}{200}\right)}} \approx 1.92$.
5. The null hypothesis cannot be rejected.

(b) 1. $H_0: p_1 = p_2$ versus $H_A: p_1 \neq p_2$.
2. $\alpha = 0.01$.
3. Reject the null hypothesis if $\chi^2 \geq 6.635$.
4. Since $\hat{p} = 0.256$, the expected frequencies for the first row are $300(0.256) = 76.8$, $200(0.256) = 51.2$; those for the second row are $300 - 76.8 = 223.2$ and $200 - 51.21 = 148.8$. Thus,
$$\chi^2 = \dfrac{(86-76.8)^2}{76.8} + \dfrac{(42-51.2)^2}{51.2} + \dfrac{(214-223.2)^2}{223.2} + \dfrac{(158-148.8)^2}{148.8} \approx 3.7.$$
5. The null hypothesis cannot be rejected.

(c) $z^2 = 1.92^2 \approx 3.69 = \chi^2$.

13.105 $400p > 5$ and $400(1-p) > 5$, so that $p > \dfrac{5}{400} = 0.0125$ and $p < 1 - \dfrac{5}{400} = 0.9875$, or $0.0125 < p < 0.9875$.

13.107 (a) $\dfrac{1.96^2}{4 \cdot 0.06^2} \approx 266.77$, which should be rounded up to 267.

(b) $\dfrac{1.96^2 \cdot 0.30 \cdot 0.70}{0.06^2} \approx 224.09$, which should be rounded up to 225.

13.109 (a) We talk about significant differences between sample data, or significant differences between sample data and population parameters, but not about significant differences between population parameters.

(b) Either delete the word "significantly" or substitute the word "sample" for the word "true."
1. $H_0: p_1 = p_2$ and $H_A: p_1 > p_2$.
2. $\alpha = 0.01$.
3. Reject the null hypothesis if $z \geq 2.33$.
4. Since $\hat{p} = \dfrac{31 + 24}{100 + 100} = 0.275$, we get $z = \dfrac{\tfrac{31}{100} - \tfrac{24}{100}}{\sqrt{(0.275)(0.725)\left(\tfrac{1}{100} + \tfrac{1}{100}\right)}} \approx 1.11$.
5. The null hypothesis cannot be rejected; the sample proportion for men is not significantly greater than that for women.

13.111 (a) $H_0: p_{11} = p_{12}, \ p_{21} = p_{22},$ and $p_{31} = p_{32}$.
$H_A: p_{i1} \neq p_{i2}$ for at least one value of i.

(b) $\alpha = 0.05$.

Reject the null hypothesis if $\chi^2 \geq 5.991$. The expected frequencies for the first row are both $\dfrac{269 \cdot 250}{500} = 134.5$, for the second row are both $\dfrac{59 \cdot 250}{500} = 29.5$, for the third row are both $250 - 134.5 - 29.5 = 86$. Thus,

$$\chi^2 = \dfrac{(146-134.5)^2}{134.5} + \dfrac{(123-134.5)^2}{134.5} + \dfrac{(28-29.5)^2}{29.5} + \dfrac{(31-29.5)^2}{29.5}$$
$$+ \dfrac{(76-86)^2}{86} + \dfrac{(96-86)^2}{86} \approx 4.445.$$

The null hypothesis cannot be rejected.

13.113 1. H_0: The manuscript is the work of the historian.
H_A: The manuscript is not the work of the historian.
2. $\alpha = 0.05$.
3. Reject the null hypothesis if $\chi^2 \geq 5.991$.
4. Since the expected frequencies are $20(0.43) = 8.6$, $20(0.32) = 6.4$, and $20(0.25) = 5.0$, we get $\chi^2 = \dfrac{(15-8.6)^2}{8.6} + \dfrac{(3-6.4)^2}{6.4} + \dfrac{(2-5.0)^2}{5.0} \approx 8.37$.
5. The null hypothesis must be rejected; the manuscript is not the work of the historian.

13.115 1. H_0: Data constitute a sample from a normal population.
H_A: Data do not constitute a sample from a normal population.
2. $\alpha = 0.05$.
3. Reject the null hypothesis if $\chi^2 \geq 5.991$.
4. Since $z = \dfrac{8.95-18.85}{5.55} \approx -1.78$, $z = \dfrac{12.95-18.85}{5.55} \approx -1.06$, $z = \dfrac{16.95-18.85}{5.55} \approx -0.34$, $z = \dfrac{20.95-18.85}{5.55} \approx 0.38$, $z = \dfrac{24.95-18.85}{5.55} \approx 1.10$, and $z = \dfrac{28.95-18.85}{5.55} \approx 1.82$, and the corresponding entries in Table I are 0.4625, 0.3554, 0.1331, 0.1480, 0.3643, and 0.4656, we find that the probabilities associated with the seven classes are $0.5000 - 0.4625 = 0.0375$, $0.4625 - 0.3554 = 0.1071$, $0.3554 - 0.1331 = 0.2223$, $0.1331 + 0.1480 = 0.2811$, $0.3643 - 0.1480 = 0.2163$, $0.4656 - 0.3643 = 0.1013$, and $0.5000 - 0.4656 = 0.0344$.

Thus, the expected frequencies are $80(0.0375) = 3.0$, $80(0.1071) \approx 8.6$, $80(0.2223) \approx 17.8$, $80(0.2811) \approx 22.5$, $80(0.2163) \approx 17.3$, $80(0.1013) \approx 8.1$, and $80(0.0344) \approx 2.8$, and, combining the first two classes and also the last two classes, we get

$$\chi^2 = \dfrac{(13-11.6)^2}{11.6} + \dfrac{(14-17.8)^2}{17.8} + \dfrac{(25-22.5)^2}{22.5} + \dfrac{(17-17.3)^2}{17.3} + \dfrac{(11-10.9)^2}{10.9} \approx 1.3.$$

5. The null hypothesis cannot be rejected; there is no real evidence that the population is not normal.

13.117 Since $\dfrac{x}{n} = \dfrac{27}{120} = 0.225$, we get $0.225 \pm 1.96 \cdot \sqrt{\dfrac{(0.225)(0.775)380}{120 \cdot 499}}$, or 0.225 ± 0.065 and $0.160 < p < 0.290$.

13.119 $C = \sqrt{\dfrac{52.8}{52.8+190}} \approx 0.47$.

13.121 With 99% confidence, the maximum error is $E = 2.575 \cdot \sqrt{\dfrac{(0.57)(0.43)}{500}} \approx 0.057$.

13.123 (a) $x = 0$ or $x \geq 9$;   (b) $x \geq 8$;
      (c) $x \leq 1$;   (d) $x \leq 1$ or $x \geq 8$.

13.125
1. H_0: $p \geq 0.30$ and H_A: $p < 0.30$.
2. $\alpha \leq 0.05$.
3. Reject the null hypothesis if $z \leq -1.645$.
4. $z = \dfrac{28 - 120(0.30)}{\sqrt{120(0.30)(0.70)}} \approx -1.59$.
5. The null hypothesis cannot be rejected; the data do not refute the claim.

13.127 In the following analysis, the deaf and blind workers are combined into one category.
1. H_0: $p_{11} = p_{12} = p_{13}$, $p_{21} = p_{22} = p_{23}$, and $p_{31} = p_{32} = p_{33}$.
   H_A: p_{i1}, p_{i2}, and p_{i3} are not all equal for at least one value of i.
2. $\alpha = 0.05$.
3. Reject the null hypothesis if $\chi^2 \geq 9.488$.
4. Since the row totals are 64, 208, and 48, and the column totals are 60, 60 and 200, the expected frequencies for the first row are $\dfrac{64 \cdot 60}{320} = 12$, 12, and $64 - (12 + 12) = 40$; those for the second row are $\dfrac{208 \cdot 60}{320} = 39$, 39, and $208 - (39 + 39) = 130$; and those for the third row are $60 - (12 + 39) = 9$, 9, and $200 - (40 + 130) = 30$. Thus,
$$\chi^2 = \dfrac{(14-12)^2}{12} + \dfrac{(14-12)^2}{12} + \dfrac{(36-40)^2}{40} + \dfrac{(35-39)^2}{39} + \dfrac{(39-39)^2}{39}$$
$$+ \dfrac{(134-130)^2}{130} + \dfrac{(11-9)^2}{9} + \dfrac{(7-9)^2}{9} + \dfrac{(30-30)^2}{30} \approx 2.49.$$
5. The null hypothesis cannot be rejected; there is no real evidence that the handicaps affect the workers' performance.

13.129 Under the null hypothesis that the coins are balanced and randomly tossed, the probabilities of the outcomes are determined by the binomial distribution with $n = 4$ and $p = 0.5$.

Number of Heads	Probability	Expected Number in 240 Trials	Observed Number in 240 Trials
0	$\frac{1}{16}$	15	32
1	$\frac{4}{16}$	60	96
2	$\frac{6}{16}$	90	62
3	$\frac{4}{16}$	60	42
4	$\frac{1}{16}$	15	8

The chi-squared statistic is $\dfrac{(32-15)^2}{15} + \dfrac{(96-60)^2}{60} + \dfrac{(62-90)^2}{90} + \dfrac{(42-60)^2}{60} + \dfrac{(8-15)^2}{15} \approx 58.24$.
This chi-squared statistic has 4 degrees of freedom. The null hypothesis should be rejected with $\alpha = 0.01$ if this exceeds $\chi^2_{0.01}$ for 4 degrees of freedom. Since $\chi^2_{0.01} = 13.277$, the null hypothesis is rejected. We decide that either the coins are not balanced or that they were not flipped properly (or both).

13.131 (a) $\dfrac{\frac{46}{102}}{\frac{28}{105}} \approx 1.69$.

    (b) 1. $H_0: p_1 = p_2$ versus $H_A: p_1 \neq p_2$.
           2. $\alpha = 0.05$.
           3. Reject the null hypothesis if $\chi^2 \geq 3.841$.
           4. The expected frequencies are $\dfrac{102 \cdot 74}{207} \approx 36.5$ and $\dfrac{102 \cdot 133}{207} \approx 65.5$ for the first row; for the second row, $\dfrac{105 \cdot 74}{207} \approx 37.5$, and $\dfrac{105 \cdot 133}{207} \approx 67.5$.

$$\chi^2 = \dfrac{(46-36.5)^2}{36.5} + \dfrac{(56-65.5)^2}{65.5} + \dfrac{(28-37.5)^2}{37.5} + \dfrac{(77-67.5)^2}{67.5} \approx 7.59.$$

           5. The null hypothesis should be rejected.

CHAPTER 14

Analysis Of Variance

14.1 For this problem $k = 3$ and $n = 5$.

(a) The three sample means (13, 17, and 15) have a variance of
$$\frac{(13-15)^2 + (17-15)^2 + (15-15)^2}{3-1} = 4$$
so that $n \cdot s_{\bar{x}}^2 = 5 \cdot 4 = 20$. The mean of the variances of the three samples is $\frac{5.5 + 18.5 + 10.0}{3} = \frac{34}{3} \approx 11.33$. Thus $F = \frac{20}{11.33} \approx 1.77$.

(b) The test at the 0.05 level of significance rejects the null hypotheses of equal manufacturer averages if the computed F exceeds 3.89, where 3.89 is the value of $F_{0.05}$ for $k-1 = 2$ and $k(n-1) = 3 \cdot (5-1) = 12$ degrees of freedom. For the actual value of 1.77, we cannot reject the null hypothesis of equal manufacturer averages.

14.3 For this problem $k = 5$ and $n = 3$.

(a) The five sample means (26.0, 24.7, 25.2, 22.7, and 26.2) have an overall mean of 24.96 and a variance of
$$\frac{(26.0-24.96)^2 + (24.7-24.96)^2 + (25.2-24.96)^2 + (22.7-24.96)^2 + (26.2-24.96)^2}{5-1}$$
$$= \frac{7.8520}{4} = 1.963$$
so that $n \cdot s_{\bar{x}}^2 = 3 \cdot 1.963 = 5.889$. The mean of the variances of the five samples is $\frac{0.37 + 1.03 + 0.63 + 0.21 + 0.21}{5} = \frac{2.45}{5} = 0.49$. Thus $F = \frac{5.889}{0.49} \approx 12.02$.

(b) The test at the 0.05 level of significance rejects the null hypotheses of equal brand averages if the computed F exceeds 3.48, where 3.48 is the value of $F_{0.05}$ for $k-1 = 4$ and $k(n-1) = 5 \cdot (3-1) = 10$ degrees of freedom. For the actual value of 12.02, we reject the null hypothesis of equal brand averages.

14.5 The standard deviations within the groups are very unequal, and this violates one of the assumptions of the method of Section 14.1.

14.7 (a) It will be impossible to distinguish the effects of the golf balls from the effects of the golf pros.

(b) It will be impossible to distinguish the effects of the golf balls from the effects of time sequence.

14.9 The randomization will be done by first assigning numbers to the persons, then selecting 5 of the 15 persons to receive diet 1, then selecting 5 of the remaining 10 persons to receive diet 2, and assigning diet 3 to the 5 persons remaining. The probability that the five heaviest persons will all get diet 1 is $\frac{1}{\binom{15}{5}} = \frac{1}{3,003}$. By the symmetry of the diets, the probability that these five persons will all get diet 2 is also $\frac{1}{3,003}$, which is also their probability of all getting diet 3. The probability that they will all get the same diet is then $\frac{3}{3,003} = \frac{1}{1,001}$, which is about 0.001.

14.11 For this problem $k = 3$ and $n = 5$.
1. H_0: $\mu_1 = \mu_2 = \mu_3$.
   H_A: The μ's are not all equal.
2. $\alpha = 0.05$.
3. Reject the null hypothesis if $F \geq 3.89$, where F is to be determined by an analysis of variance and 3.89 is the value of $F_{0.05}$ for $k - 1 = 2$ and $k(n - 1) = 12$ degrees of freedom; otherwise, accept the null hypothesis or reserve judgment.
4. Substituting the given values into the computing formulas, we get
$$SST = 3,551 - \frac{1}{15}(225)^2 = 176$$
$$SS(Tr) = \frac{1}{5}(65^2 + 85^2 + 75^2) - \frac{1}{15}(225)^2 = 40$$
$$SSE = 176 - 40 = 136$$
We can summarize the results in the following analysis of variance table:

Source of variation	Degrees of freedom	Sum of squares	Mean square	F
Treatments	2	40	20.00	1.77
Error	12	136	11.33	
Total	14	176		

5. Since $F = 1.77$ does not exceed 3.89, the null hypothesis cannot be rejected.

14.13 For this problem $k = 5$ and $n = 3$.
1. H_0: $\mu_1 = \mu_2 = \mu_3 = \mu_4 = \mu_5$.
   H_A: The μ's are not all equal.
2. $\alpha = 0.05$.
3. Reject the null hypothesis if $F \geq 3.48$, where F is to be determined by an analysis of variance and 3.48 is the value of $F_{0.05}$ for $k - 1 = 4$ and $k(n - 1) = 10$ degrees of freedom; otherwise, accept the null hypothesis or reserve judgment.
4. Substituting the given values into the computing formulas, we get
$$SST = 9,373.48 - \frac{1}{15}(374.4)^2 = 28.456$$
$$SS(Tr) = \frac{1}{3}(78.0^2 + 74.1^2 + 75.6^2 + 68.1^2 + 78.6^2) - \frac{1}{15}(374.4)^2 = 23.556$$
$$SSE = 28.456 - 23.556 = 4.900$$
We can summarize the results in the following analysis of variance table:

Source of variation	Degrees of freedom	Sum of squares	Mean square	F
Treatments	4	23.556	5.889	12.02
Error	10	4.900	0.490	
Total	19	28.456		

5. Since $F = 12.02$ exceeds 3.48, the null hypothesis must be rejected.

Modern Elementary Statistics – 9th Edition   93

14.15 1. $H_0: \mu_1 = \mu_2 = \mu_3 = \mu_4$ and
             H_A: The μ's are not all equal.
         2. $\alpha = 0.05$.
         3. Reject the null hypothesis if $F \geq 3.24$.
         4. Since $k = 4$, $n = 5$, $T_{1.} = 55$, $T_{2.} = 60$, $T_{3.} = 64$, $T_{4.} = 54$, $T_{..} = 233$, and $\Sigma x^2 = 2{,}829$, we get $SST = 2{,}829 - \frac{1}{20}(233)^2 = 114.55$, $SS(Tr) = \frac{1}{5}(13{,}637) - 2{,}714.45 = 12.95$,
             $SSE = 114.55 - 12.95 = 101.60$, $MS(Tr) = \frac{12.95}{3} \approx 4.32$, $MSE = \frac{101.60}{16} \approx 6.35$, and
             $F = \frac{4.32}{6.35} \approx 0.68$.
         5. The null hypothesis cannot be rejected; there is no real indication that there is a difference in the numbers of mistakes averaged by the four compositors.

14.17 1. $H_0: \mu_G = \mu_B = \mu_J = \mu_M$ versus
             H_A: The μ's are not all equal.
         2. $\alpha = 0.01$.
         3. Reject the null hypothesis if $F \geq 5.01$, where F is to be determined by an analysis of variance and 5.01 is the value of $F_{0.01}$ for 3 and 19 degrees of freedom. Otherwise accept the null hypothesis or reserve judgment.
         4. The group totals are $T_G = 379$, $T_H = 360$, $T_J = 276$, and $T_M = 445$, with grand total 1,460 and with $\sum_{i=1}^{4}\sum_{j=1}^{n_i} x_{ij}^2 = 104{,}500$. This leads to $SST = 104{,}500 - \frac{1{,}460^2}{23} \approx 11{,}821.74$.
             It follows also that $SS(Tr) = \frac{379^2}{6} + \frac{360^2}{5} + \frac{276^2}{7} + \frac{445^2}{5} - \frac{1{,}460^2}{23} \approx 7{,}669.20$. This leads to the following analysis of variance:

Source of variation	Degrees of freedom	Sum of squares	Mean square	F
Models	3	7,669.20	2,556.40	11.70
Error	19	4,152.54	218.55	
Total	22	11,821.74		

         5. Since $11.70 > 5.01$, we reject the null hypothesis and conclude that the models are significantly different.

14.19 1. $H_0: \mu_1 = \mu_2 = \mu_3$ and
             H_A: The μ's are not all equal.
         2. $\alpha = 0.05$.
         3. Reject the null hypothesis if $F \geq 3.89$.
         4. Using the condition totals 83, 138, and 133, with a grand total of 354, and with sum of squares 8,584, we find $SST = 8{,}584 - \frac{354^2}{15} \approx 229.60$,
             $SS(Tr) = \frac{83^2}{4} + \frac{138^2}{6} + \frac{133^2}{5} - \frac{354^2}{15} \approx 79.65$, then $SSE = 229.60 - 79.65 = 149.95$. This leads to the following analysis of variance:

Source of variation	Degrees of freedom	Sum of squares	Mean square	F
Treatments	2	79.65	39.82	3.19
Error	12	149.95	12.50	
Total	14	229.60		

         5. The null hypothesis cannot be rejected; the differences among the three sample means are not significant.

14.21 $(k-1) + k(n-1) = k - 1 + kn - k = kn - 1$.

14.23 (a) $\dfrac{SS(Tr)}{k-1} = \dfrac{n \cdot \sum_{i=1}^{k}(\bar{x}_{i.} - \bar{x}_{..})^2}{k-1} = n \cdot \left[\dfrac{\sum_{i=1}^{k}(\bar{x}_{i.} - \bar{x}_{..})^2}{k-1}\right] = n \cdot s_{\bar{x}}^2$.

(b) $\dfrac{SSE}{k(n-1)} = \dfrac{\sum_{i=1}^{k}\sum_{j=1}^{n}(x_{ij} - \bar{x}_{i.})^2}{k(n-1)} = \dfrac{1}{k} \cdot \sum_{i=1}^{k}\left[\dfrac{\sum_{j=1}^{n}(x_{ij} - \bar{x}_{i.})^2}{n-1}\right] = \dfrac{1}{k}\sum_{i=1}^{k} s_i^2$.

14.25 (a) The first step is the completion of the analysis of variance. Note that $k = 8$ and $n = 6$. The analysis of variance table is

Source of variation	Degrees of freedom	Sum of squares	Mean square	F
Treatments	7	12,696.20	1,813.74	9.28
Error	40	7,818.70	195.47	
Total	47	20,514.90		

Since $F_{0.05} = 2.25$ for 7 and 40 degrees of freedom, and since $9.28 > 2.25$, we reject the null hypothesis of equal treatment means.

(b) The Studentized range method uses $q_{0.05}$ with $k = 8$ treatments and $k(n-1) = 40$ degrees of freedom. Table XII indicates that $q_{0.05} = 4.52$. Observe next that $s = \sqrt{195.47} \approx 13.98$. The standard for comparing two treatments is $\dfrac{4.52}{\sqrt{n}} \cdot s = \dfrac{4.52}{\sqrt{8}} \cdot 13.98 \approx 25.80$. Now sort the spraying schedules by their means:

80.17	104.00	111.17	120.33	121.83	127.50	129.17	133.50
D	E	H	A	C	G	F	B

We can find the following pairs significantly different:
D vs H
D vs A
D vs C
D vs G
D vs F
D vs B   E vs B

14.27 She might consider only programs of the same length, or she might use the program lengths as blocks and perform a two-way analysis of variance.

14.29 1. H_0's: $\alpha_1 = \alpha_2 = \alpha_3 = 0$: $\beta_1 = \beta_2 = \beta_3 = \beta_4 = 0$.
H_A's: The treatment effects are not all equal to zero; the block effects are not all equal to zero.
2. $\alpha = 0.05$ for both tests.
3. For treatments, reject the null hypothesis if $F \geq 5.14$; for blocks, reject the null hypothesis if $F \geq 4.76$.
4. Since $k = 3$, $n = 4$, $T_{1.} = 13.0$, $T_{2.} = 11.4$, $T_{3.} = 13.2$, $T_{.1} = 10.3$, $T_{.2} = 8.8$, $T_{.3} = 8.8$, $T_{.4} = 9.7$, $T_{..} = 37.6$, and $\Sigma x^2 = 119.06$, we get $SST = 119.06 - \frac{1}{12}(37.6)^2 \approx 1.25$, $SS(Tr) = \frac{1}{4}(473.2) - 117.81 = 0.49$, and $SSB = \frac{1}{3}(355.06) - 117.81 \approx 0.54$, $SSE = 1.25 - 0.49 - 0.54 = 0.22$, $MS(Tr) = \frac{0.49}{2} = 0.245$, $MSB = \frac{0.54}{3} = 0.18$, $MSE = \frac{0.22}{6} \approx 0.037$, and $F = \frac{0.245}{0.037} \approx 6.62$ for treatments and $F = \frac{0.180}{0.037} \approx 4.86$ for blocks.
5. Both null hypotheses must be rejected; the differences among the sample means obtained for the three diet foods are significant and so are the differences among the sample means obtained for the four laboratories.

14.31 1. H_0's: $\alpha_1 = \alpha_2 = \alpha_3 = \alpha_4 = \alpha_5 = 0$: $\beta_1 = \beta_2 = \beta_3 = \beta_4 = 0$.
H_A's: The treatment effects are not all equal to zero; the block effects are not all equal to zero.
2. $\alpha = 0.05$ for both tests.
3. For treatments, reject the null hypothesis if $F \geq 3.26$; for blocks, reject the null hypothesis $F \geq 3.49$.
4. Since $k = 5$, $n = 4$, $T_{1.} = 83.1$, $T_{2.} = 103.0$, $T_{3.} = 94.5$, $T_{4.} = 95.2$, $T_{5.} = 84.0$, $T_{.1} = 115.8$, $T_{.2} = 112.1$, $T_{.3} = 114.0$, $T_{.4} = 117.9$, $T_{..} = 459.8$, and $\Sigma x^2 = 10,669.98$, we get $SST = 10,669.98 - \frac{1}{20}(459.8)^2 = 99.18$, $SS(Tr) = \frac{1}{4}(42,563.9) - 10,570.80 = 70.18$, $SSB = \frac{1}{5}(52,872.46) - 10,570.8 = 3.69$, $SSE = 99.18 - 70.18 - 3.69 = 25.31$, $MS(Tr) = \frac{70.18}{4} = 17.54$, $MSB = \frac{3.69}{3} = 1.23$, $MSE = \frac{25.31}{12} \approx 2.11$, and $F = \frac{17.54}{2.11} \approx 8.31$ for treatments and $F = \frac{1.23}{2.11} \approx 0.58$ for blocks.
5. The null hypothesis for treatments must be rejected; the null hypothesis for blocks cannot be rejected.

14.33 1. H_0's: $\alpha_1 = \alpha_2 = \alpha_3 = 0$ and $\beta_1 = \beta_2 = \beta_3 = \beta_4 = 0$.
  H_A's: The α's are not all equal to zero; the β's are not all equal to zero.
2. $\alpha = 0.05$ for both tests.
3. For factor A (launchers), reject the null hypothesis if $F \geq 5.14$; for factor B (fuels), reject the null hypothesis if $F \geq 4.76$.
4. Since $k = 3$, $n = 4$, $T_{1\cdot} = 197.4$, $T_{2\cdot} = 185.9$, $T_{3\cdot} = 206.0$, $T_{\cdot 1} = 137.6$, $T_{\cdot 2} = 165.5$, $T_{\cdot 3} = 157.6$, $T_{\cdot 4} = 128.6$, $T_{\cdot\cdot} = 589.3$, and $\Sigma x^2 = 29{,}318.55$, we get
$SST = 29{,}318.55 - \frac{1}{12}(589.3)^2 \approx 379.01$, $SSA = 28{,}990.39 - 28{,}939.54 = 50.85$,
$SSB = 29{,}233.24 - 28{,}939.54 = 293.70$, $SSE = 379.01 - 50.85 - 293.70 = 34.46$,
$MSA = \frac{50.85}{2} = 25.42$, $MSB = \frac{293.70}{3} = 97.90$, $MSE = \frac{34.46}{6} \approx 5.74$, and
$F = \frac{25.42}{5.74} \approx 4.43$ for factor A and $F = \frac{97.90}{5.74} \approx 17.06$ for factor B.
5. The null hypothesis for factor A cannot be rejected; the null hypothesis for factor B must be rejected. We conclude that the fuels, but not the launchers, affect the range of the missile.

14.35 (a) 1 and I, 2 and I, 3 and I, 4 and I, 5 and I, 1 and II, 2 and II, 3 and II, 4 and II, 5 and II, 1 and III, 2 and III, 3 and III, 4 and III, and 5 and III.
(b) $4 \cdot 5 \cdot 3 = 60$.

14.37 $A_L B_L C_L D_L$, $A_L B_L C_L D_H$, $A_L B_L C_H D_L$, $A_L B_L C_H D_H$, $A_L B_H C_L D_L$, $A_L B_H C_L D_H$, $A_L B_H C_H D_L$, $A_L B_H C_H D_H$, $A_H B_L C_L D_L$, $A_H B_L C_L D_H$, $A_H B_L C_H D_L$, $A_H B_L C_H D_H$, $A_H B_H C_L D_L$, $A_H B_H C_L D_H$, $A_H B_H C_H D_L$, $A_H B_H C_H D_H$.

14.39 1. H_0's: The row effects are all equal to zero; the column effects are all equal to zero; the treatment effects are all equal to zero.
  H_A's: The row effects are not all equal to zero; the column effects are not all equal to zero; the treatment effects are not all equal to zero.
2. $\alpha = 0.05$ for each test.
3. For rows, columns, or treatments, reject the null hypothesis if $F \geq 19.0$.
4. Since $r = 3$, $T_{1\cdot} = 230$, $T_{2\cdot} = 260$, $T_{3\cdot} = 246$, $T_{\cdot 1} = 240$, $T_{\cdot 2} = 248$, $T_{\cdot 3} = 248$, $T_A = 244$, $T_B = 274$, $T_C = 218$, $T_{\cdot\cdot} = 736$, and $\Sigma x^2 = 60{,}882$, we get
$SST = 60{,}882 - \frac{1}{9}(736)^2 \approx 693.56$, $SSR = \frac{1}{3}(181{,}016) - 60{,}188.44 \approx 150.23$,
$SSC = \frac{1}{3}(180{,}608) - 60{,}188.44 \approx 14.23$, $SS(Tr) = \frac{1}{3}(182{,}136) - 60{,}188.44 = 523.56$,
$SSE = 693.56 - 150.23 - 14.23 - 523.56 = 5.54$,
$MSR = \frac{150.23}{2} = 75.12$, $MSC = \frac{14.23}{2} \approx 7.12$, $MS(Tr) = \frac{523.56}{2} = 261.78$,
$MSE = \frac{5.54}{2} \approx 2.77$, and $F = \frac{75.12}{2.77} \approx 27.12$ for rows, $F = \frac{7.12}{2.79} \approx 2.57$ for columns, and
$F = \frac{261.78}{2.77} \approx 94.51$ for treatments.
5. The null hypothesis for rows (professional interest) must be rejected; the null hypothesis for columns (ethnic background) cannot be rejected; the null hypothesis for treatments (instructors) must be rejected. We conclude that professional interests and instructors, but not ethnic backgrounds, affect the scores.

14.41 Letting I stand for Independent, R for Republican, and D for Democrat, we can display the given information as follows:

	Teacher	Lawyer	Doctor
Easterner	I		
Southerner	R	D	
Westerner			

Completing the Latin Square, which is not difficult, leads to the result that the doctor who is a Westerner is a Republican.

14.43 Since 2 already appears together with 1, 3, 4, and 6, it must appear together with 5 and 7 on Thursday.
Since 4 already appears together with 1, 2, 5, and 6, it must appear together with 3 and 7 on Tuesday.
Since 5 already appears together with 1, 2, 4, and 7, it must appear together with 3 and 6 on Saturday.

14.45 (a) Since the means are $\bar{x}_1 = \frac{44}{4} = 11$, $\bar{x}_2 = \frac{60}{4} = 15$, and $\bar{x}_3 = \frac{40}{4} = 10$, and their mean is $\frac{11+15+10}{3} = 12$, we get $n \cdot s_{\bar{x}}^2 = 4 \cdot \frac{(11-12)^2 + (15-12)^2 + (10-12)^2}{2} = 28$. Also, since $s_1^2 = \frac{26}{3}$, $s_2^2 = \frac{34}{3}$, and $s_3^2 = \frac{26}{3}$, their mean is $\frac{86}{9} \approx 9.56$, and $F = \frac{28}{9.56} \approx 2.93$.

(b) 1. H_0: $\mu_1 = \mu_2 = \mu_3$ and H_A: The μ's are not all equal.
2. $\alpha = 0.01$.
3. Reject the null hypothesis if $F \geq 8.02$.
4. $F = 2.93$ from part (a).
5. The null hypothesis cannot be rejected; the differences among the three sample means can be attributed to chance.

14.47 (a) It is a balanced incomplete block design because the seven department heads are not all serving together on a committee, but each department head serves together with each other department head on two committees.
(b) There are two solutions:
Griffith -- Dramatics
Anderson -- Discipline
Evans -- Tenure or salaries
Fleming -- Salaries or tenure

14.49 1. H_0's: $\alpha_1 = \alpha_2 = \alpha_3 = 0$; $\beta_1 = \beta_2 = \beta_3 = 0$
H_A's: The α's are not all equal to zero; the β's are not all equal to zero.
2. $\alpha = 0.05$ for both tests.
3. For factor A, reject the null hypothesis if $F \geq 6.94$; for factor B reject the null hypothesis if $F \geq 6.94$.
4. Since $k = 3$, $n = 3$, $T_{1.} = 236$, $T_{2.} = 266$, $T_{3.} = 252$, $T_{.1} = 250$, $T_{.2} = 280$, $T_{.3} = 224$, $T_{..} = 754$, and $\Sigma x^2 = 63{,}862$, we get $SST = 63{,}862 - \frac{1}{9}(754)^2 \approx 693.56$,

$SSA = \frac{1}{3}(189{,}956) - 63{,}168.44 \approx 150.23$,

$SSB = \frac{1}{3}(191{,}076) - 63{,}168.44 = 523.56$, $SSE = 693.56 - 150.23 - 523.56 = 19.77$,

$MSA = \frac{150.23}{2} \approx 75.12$, $MSB = \frac{523.56}{2} = 261.78$, $MSE = \frac{19.77}{4} \approx 4.94$, and

$F = \frac{75.12}{4.94} \approx 15.21$ for factor A and $F = \frac{261.78}{4.94} \approx 52.99$ for factor B.
5. The null hypothesis for factor A (majors) must be rejected; the null hypothesis for factor B (instructors) must be rejected.

14.51 1. H_0: $\mu_1 = \mu_2 = \mu_3$ and H_A: The μ's are not all equal.
2. $\alpha = 0.05$.
3. Reject the null hypothesis if $F \geq 4.26$.
4. Since $n_1 = 4$, $n_2 = 5$, $n_3 = 3$, $N = 12$, $T_{1.} = 332$, $T_{2.} = 535$, $T_{3.} = 285$, $T_{..} = 1{,}152$, and $\Sigma x^2 = 112{,}722$, we get $SST = 112{,}722 - \frac{1}{12}(1{,}152)^2 = 2{,}130$,

$SS(Tr) = 111{,}876 - 110{,}592 = 1{,}284$, $SSE = 2{,}130 - 1{,}284 = 846$, $MS(Tr) = \frac{1{,}284}{2} = 642$,

$MSE = \frac{846}{9} = 94$, and $F = \frac{642}{94} \approx 6.83$.
5. The null hypothesis must be rejected; there is real evidence that the different descriptions affect sales.

14.53 The result is attributed to rank.

14.55 H_0's: $\alpha_1 = \alpha_2 = \alpha_3 = \alpha_4 = 0$; $\beta_1 = \beta_2 = \beta_3 = \beta_4 = \beta_5 = 0$
H_A's: The α's are not all equal to zero; the β's are not all equal to zero.
2. $\alpha = 0.05$ for both tests.
3. For factor A, reject the null hypothesis if $F \geq 3.49$; for blocks reject the null hypothesis if $F \geq 3.46$.
4. Since $k = 4$, $n = 5$, $T_{1.} = 134$, $T_{2.} = 140$, $T_{3.} = 156$, $T_{4.} = 146$, $T_{.1} = 114$, $T_{.2} = 117$, $T_{.3} = 115$, $T_{.4} = 127$, $T_{.5} = 103$, $T_{..} = 576$, and $\Sigma x^2 = 16{,}742$, we get

$SST = 16{,}742 - \frac{1}{20}(576)^2 = 153.2$, $SS(Tr) = 16{,}641.6 - 16{,}588.8 = 52.8$,

$SSB = 16{,}662 - 16{,}588.8 = 73.2$, $SSE = 153.2 - 52.8 - 73.2 = 27.2$,

$MS(Tr) = \frac{52.8}{3} = 17.6$, $MSB = \frac{73.2}{4} = 18.3$, $MSE = \frac{27.2}{12} = 2.27$, and $F = \frac{17.6}{2.27} \approx 7.75$

for treatments and $F = \frac{18.3}{2.27} \approx 8.06$ for blocks.
5. The null hypothesis for treatments (routes) must be rejected; the null hypothesis for blocks (days of the week) must be rejected.

CHAPTER 15

Regression

15.1 For the first line we get $y = 10 - \frac{1}{2} \cdot 6 = 7$, $y = 10 - \frac{1}{2} \cdot 12 = 4$, and $y = 10 - \frac{1}{2} \cdot 18 = 1$, so that the sum of the squares of the errors is $(5-7)^2 + (6-4)^2 + (1-1)^2 = 8$. For the second line we get $y = 8 - \frac{1}{3} \cdot 6 = 6$, $y = 8 - \frac{1}{3} \cdot 12 = 4$, and $y = 8 - \frac{1}{3} \cdot 18 = 2$, so that the sum of the squares of the errors is $(5-6)^2 + (6-4)^2 + (1-2)^2 = 6$. Thus, the second line provides a better fit.

15.3 (a) The normal equations are
$107 = 6a + 36b$
$721 = 36a + 304b$
By the method of elimination
$642 = 36a + 216b$
$721 = 36a + 304b$
$\,79 = \,88b$
so that $b = \frac{79}{88} \approx 0.898$ and $a = \frac{107 - 36(0.898)}{6} \approx 12.445$.

(b) $12.445 + 0.898(8) \approx 19.6$;

(c) The prediction for someone who has been at the inspection station for 120 weeks is $12.445 + 0.989(120) \approx 131.1$. This is of course ridiculous, but this is what happens when we make predictions for values very different from those in the original database.

15.5 The conclusion does not follow; the relationship applies only to the range of values of x (prices) on which the study is based. Since the prices ranged from \$1.50 to \$2.40, we cannot extrapolate for \$0.10.

15.7 Since $n = 8$, $\Sigma x = 98.1$, $\Sigma x^2 = 1,299.85$, $\Sigma y = 435.8$, and $\Sigma xy = 5,772.65$, we get
$S_{xx} = 1,299.85 - \frac{1}{8}(98.1)^2 \approx 96.90$, $S_{xy} = 5,772.65 - \frac{1}{8}(98.1)(435.8) = 428.65$,
$b = \frac{428.65}{96.90} \approx 4.424$, and $a = \frac{435.8 - 4.424(98.1)}{8} \approx 0.23$. Thus, the equation of the least-squares line is $\hat{y} = 0.23 + 4.42x$.

15.9 No. This does not affect their actual income.

15.11 There is no variation in the x's, and she will find $S_{xx} = 0$. This makes it impossible to determine the regression slope.

15.13 Denoting the missing value by y, we get $n = 3$, $\Sigma x = 306$, $\Sigma y = 192 + y$, so that substitution into the first of the two normal equations yields
$192 + y = 3a + 306b$
$ = 3(48)306(0.5) = 297$
and $y = 297 - 192 = 105$.

15.15 Since $n = 12$, $\Sigma x = 507$, $\Sigma y = 144$, $\Sigma y^2 = 1{,}802$, $\Sigma xy = 6{,}314$ we get $S_{yy} = 1{,}802 - \frac{1}{12}(144)^2 = 74$, $S_{xy} = 6{,}314 - \frac{1}{12}(507)(144) = 230$, $b = \frac{230}{74} \approx 3.108$, and $a = \frac{507 - 3.108(144)}{12} \approx 4.954$. The equation of the least-squares line is $\hat{x} = 4.954 + 3.108y$. For $y = 10$, the predicted value is $\hat{x} = 4.954 + 3.108(10) \approx 36.03$.

15.17 (a) We get $S_{xx} = 385 - \frac{1}{10}(55)^2 = 82.5$, $S_{xy} = 76 - \frac{1}{10}(0)^2 = 76$, $b = \frac{76}{82.5} \approx 0.92$, and $a = \frac{0 - 0.921(55)}{10} \approx -5.07$. The equation of the least-squares line is $\hat{y} = -5.07 + 0.92x$.

(b) For $x = 1$, $\hat{y} = -5.07 + 0.92(1) = -4.15$; for $x = 10$, $\hat{y} = -5.07 + 0.92(10) = 4.13$.

(c) If the SAT ranks were assigned to the ten students at random, the expected rank for each student would be 5.5; the student ranked 1 on the PSAT test could expect his ranking to go down 4.5 points and the one ranked 10 could expect his ranking to go up 4.5 points.

15.19
1. $H_0: \beta = 1.2$ and $H_A: \beta < 1.2$.
2. $\alpha = 0.05$.
3. Reject the null hypothesis if $t \leq -2.132$.
4. Using values given in Exercise 15.3, we get $S_{xx} = 304 - \frac{1}{6}(36)^2 = 88$, $S_{xy} = 721 - \frac{1}{6}(36)(107) = 79$, and $S_{yy} = 2{,}001 - \frac{1}{6}(107)^2 \approx 92.83$. Thus $b = \frac{79}{88} \approx 0.898$, $s_e = \sqrt{\frac{92.83 - 0.898(79)}{4}} \approx 2.34$, and $t = \frac{0.898 - 1.2}{\frac{2.34}{\sqrt{88}}} \approx -1.21$.
5. The null hypothesis cannot be rejected.

15.21 (a) $\hat{y} = 106.026 + 14.927x$.

(b)
1. $H_0: \beta = 10.3$ versus $H_A: \beta > 10.3$.
2. $\alpha = 0.05$.
3. Reject the null hypothesis if $t \geq 1.943$, where $t = \frac{b - 10.3}{\frac{s_e}{\sqrt{S_{xx}}}}$ and 1.943 is the value of $t_{0.05}$ on 6 degrees of freedom.
4. As a by-product of the regression, we have $S_{xx} = 23.8348$, $S_{xy} = 355.7910$, $S_{yy} = 5{,}356.5938$, and $s_e = 2.7588$. Then we find $t = \frac{14.927 - 10.3}{\frac{2.7558}{\sqrt{23.8348}}} \approx 8.197$.
5. The null hypothesis must be rejected.

15.23
1. $H_0: \alpha = 2.1$ and $H_A: \alpha < 2.1$.
2. $\alpha = 0.05$.
3. Reject the null hypothesis if $t \leq -2.132$.
4. Since $\bar{x} = \frac{42}{6} = 7$, we get $t = \frac{1.902 - 2.1}{0.077 \cdot \sqrt{\frac{1}{6} + \frac{7^2}{70}}} \approx -2.76$.
5. The null hypothesis must be rejected.

Modern Elementary Statistics – 9th Edition   101

15.25 (a) We get $S_{xx} = 1{,}166 - \frac{1}{6}(82)^2 \approx 45.33$, $S_{xy} = 4{,}579 - \frac{1}{6}(82)(382) \approx -641.67$,

$b = \frac{-641.67}{45.33} \approx -14.16$, $a = \frac{382 - (-14.16)(82)}{6} \approx 257.19$, and $\hat{y} = 257.19 - 14.16x$.

(b) $s_e = \sqrt{\frac{9{,}459.33 - (-14.16)(-641.67)}{4}} \approx 9.66$, and the interval is $-14.16 \pm 2.776 \cdot \frac{9.66}{\sqrt{45.33}}$,

or -14.16 ± 3.98, and $-18.14 < \beta < -10.18$.

15.27 $1.483 \pm 3.747 \frac{0.16}{\sqrt{3.44}}$, or 1.483 ± 0.323; $1.160 < \beta < 1.806$. We are estimating the average increase in advertising expenses corresponding to a unit increase in operating profits.

15.29 $1.259 \pm 4.604(0.16) \cdot \sqrt{\frac{1}{6} + \frac{1.5^2}{3.44}}$, or 1.259 ± 0.667; $0.592 < \alpha < 1.926$.

15.31 Using $S_{yy} = 92.8333$, we get $s_e^2 = \frac{92.83333 - \frac{79^2}{88}}{6 - 2} \approx 5.4782$. Then $s_e \approx 2.341$. The point prediction for a person with eight weeks of experience is $12.455 + 0.898(8) = 19.629$. The 95 percent confidence interval is then $19.629 \pm 2.776 \cdot 2.341 \sqrt{\frac{1}{6} + \frac{(8-6)^2}{88}}$ where 2.776 is $t_{0.025}$ for $n - 2 = 4$ degrees of freedom. This interval can be written as 19.629 ± 2.993. This can be rounded to $(16.6, 22.6)$.

15.33 Since $\hat{y} = 1.259 + 1.438(2) = 4.225$, we get $4.225 \pm 2.776(0.16) \cdot \sqrt{\frac{1}{6} + \frac{(2-1.5)^2}{3.44}}$, or 4.225 ± 0.217; $4.01 < \mu_{y|2} < 4.44$.

15.35 The procedure is identical to that of problem 15.32, except that we are making a prediction for a single house (rather than estimating the mean price of houses assessed at $4,500). The prediction interval is $173.20 \pm 2.447 \cdot 2.7558 \sqrt{1 + \frac{1}{8} + \frac{(4.5 - 5.0938)^2}{23.8384}}$ or 173.20 ± 7.20. This interval is $(166.00, 180.40)$.

15.37 $s_e^2 = \frac{S_{xx} \cdot S_{yy} - (S_{xy})^2}{(n-2)S_{xx}} = \frac{S_{yy} - \frac{(S_{xy})^2}{S_{xx}}}{n-2} = \frac{S_{yy} - \frac{S_{xy}}{S_{xx}} \cdot S_{xy}}{n-2} = \frac{S_{yy} - b \cdot S_{xy}}{n-2}$.

15.39 Since $n = 6$, $\Sigma x_1 = 18.0$, $\Sigma x_2 = 9.1$, $\Sigma x_1^2 = 63.6$, $\Sigma x_1 x_2 = 27.0$, $\Sigma x_2^2 = 15.29$, $\Sigma y = 740$, $\Sigma x_1 y = 2{,}505.4$, and $\Sigma x_2 y = 1{,}131.4$, we get
$740.0 = 6.0 b_0 + 18.0 b_1 + 9.10 b_2$
$2{,}505.4 = 18.0 b_0 + 63.6 b_1 + 27.00 b_2$
$1{,}131.4 = 9.1 b_0 + 27.0 b_1 + 15.29 b_2$
The solutions of these normal equations are $b_0 = 14.56$, $b_1 = 30.109$, and $b_2 = 12.16$.
(a) $\hat{y} = 14.56 + 30.11 x_1 + 12.16 x_2$;
(b) $14.56 + 30.11(2.4) + 12.16(1.2) \approx \$101.42$.
(c) $14.56 + 30.11(2.4) + 12.16(1.5) \approx \$105.06$.
(d) $14.56 + 30.11(3.6) + 12.16(1.2) \approx \$137.55$.
(e) $14.56 + 30.11(3.6) + 12.16(1.5) \approx \$141.20$.

15.41 Since $n=6$, $\Sigma x_1 = 0$, $\Sigma x_2 = 0$, $\Sigma x_1^2 = 4$, $\Sigma x_1 x_2 = 0$, $\Sigma x_2^2 = 6$, $\Sigma y = 418.4$, $\Sigma x_1 y = 11.9$, and $\Sigma x_2 y = -71.8$, we get $b_0 = \dfrac{418.4}{6} \approx 69.73$, $b_1 = \dfrac{11.9}{4} = 2.975$, and $b_2 = \dfrac{-71.8}{6} \approx -11.97$.

(a) $\hat{y} = 69.73 + 2.975 x_1 - 11.97 x_2$.

(b) Substituting $x_1 = 0.5$ and $x_2 = 0$, we get $\hat{y} = 69.73 + 2.975(0.5) - 11.97(0) \approx 71.2$. The difference is due to rounding.

15.43 Since $n = 5$, $\Sigma x = 30$, $\Sigma x^2 = 220$, $\Sigma \log y = 11.9286$, and $\Sigma x(\log y) = 75.2228$, we get
$$11.9286 = 5(\log a) + 30(\log b)$$
$$75.2228 = 30(\log a) + 220(\log b)$$
and the solutions of these normal equations are $\log a = 1.8380$ and $\log b = 0.0913$, and $a = 68.9$ and $b = 1.23$. Thus, the equation of the exponential curve is $\hat{y} = 68.9(1.23)^x$. Also, for $x = 5$ we get $\log \hat{y} = 1.8380 + (0.0913)5 = 2.2945$ and $\hat{y} = 197$ thousand.

15.45 Since $n = 5$, $\Sigma \log x = 5.6926$, $\Sigma (\log x)^2 = 6.5405$, $\Sigma \log y = 8.7320$ and $\Sigma (\log x)(\log y) = 9.8014$, we get
$$8.7320 = 5(\log a) + 5.6926 b$$
$$9.8014 = 5.6926(\log a) + 6.5405 b$$
and the solutions of these normal equations are $\log a = 4.4347$ and $b = -2.36$, and $a = 27,200$. Thus, the equation of the power function is $\hat{y} = 27,200 \cdot x^{-2.36}$. Also, for $x = 12$ we get $\log \hat{y} = 4.4347 - 2.36(\log 12) \approx 1.8878$ and $\hat{y} = 77.2$ thousand units.

15.47 Since $n = 6$, $\Sigma x = 7.6$, $\Sigma x^2 = 12.90$, $\Sigma x^3 = 24.208$, $\Sigma x^4 = 47.5074$, $\Sigma y = 157.8$, $\Sigma xy = 189.03$ and $\Sigma x^2 y = 298.083$, we get
$$157.8 = 6.0 b_0 + 7.6 b_1 + 12.90 b_2$$
$$189.03 = 7.6 b_0 + 12.90 b_1 + 24.208 b_2$$
$$298.083 = 12.90 b_0 + 24.208 b_1 + 47.5074 b_2$$
and the solutions of these normal equations are $b_0 = 14.789$, $b_1 = 38.892$, and $b_2 = -17.559$. Thus, the equation of the parabola is $\hat{y} = 14.789 + 38.892 x - 17.559 x^2$. Also, for $x = 1.5$ we get $\hat{y} = 14.789 + 38.892(1.5) - 17.559(1.5)^2 \approx 33.6$ pounds per square yard.

15.49 Since $n = 7$, $\Sigma x = 0$, $\Sigma x^2 = 28$, $\Sigma x^3 = 0$, $\Sigma x^4 = 196$, $\Sigma y = 23.6$, $\Sigma xy = 4.2$ and $\Sigma x^2 y = 99.2$, we get
$$23.6 = 7.0 b_0 + 28 b_2$$
$$4.2 = 28 b_1$$
$$99.2 = 28 b_0 + 196 b_2$$
and the solutions of these normal equations are $b_0 = 3.143$, $b_1 = 0.15$, and $b_2 = 0.057$. Thus, the equation is $\hat{y} = 3.143 + 0.15 x + 0.057 x^2$.

15.51 This person is guilty of using a regression to make a prediction outside the experience of the regression. The regression was done on prices over the range 50 cents to 65 cents and can give bad results if used at 85 cents. The database, in fact, showed sales declining as prices rose.
The parabola cannot be used for values of x beyond the range of values of x investigated, because it turns up again for larger values of x.

Modern Elementary Statistics – 9th Edition 103

15.53 We get $S_{yy} = 374 - \frac{1}{10}(54)^2 = 82.4$, $s_e = \sqrt{\frac{82.4-(-0.3326)167}{8}} \approx 2.446$, and the interval is $-0.3326 \pm 2.306 \cdot \frac{2.446}{\sqrt{312.1}}$, or -0.3326 ± 0.3193, or $-0.652 < \beta < -0.013$.

15.55 Use $-2, -1, 0, 1$, and 2 as the month numbers. Then the data looks like this:

x_1	$x_2 = x_1^2$	y
-2	4	46
-1	1	55
0	0	57
1	1	52
2	4	44

You can find easily that $\Sigma x_1 = 0$, $\Sigma x_1^2 = 10 = \Sigma x_2$, $\Sigma x_1 x_2 = 0$, $\Sigma x_2^2 = 34$, $\Sigma y = 254$, $\Sigma x_1 y = -51$, and $\Sigma x_2 y = 467$. You can construct the normal equations as
$254 = 5b_0 + 0b_1 + 10b_2$
$-51 = 0b_0 + 10b_1 + 0b_2$
$467 = 10b_0 + 0b_1 + 34b_2$
The middle equation solves as $b_1 = -5.1$. The outer equations are:
$254 = 5b_0 + 10b_2$
$467 = 10b_0 + 34b_2$
and they can be solved as $b_2 = -2.93$ and $b_0 = 56.66$. The fitted equation is $\hat{y} = 56.66 - 5.1x_1 - 2.93x_2$.

15.57 Since $n = 6$, $\Sigma x = 125.8$, $\Sigma x^2 = 5{,}338.34$, $\Sigma \log y = 9.5879$, and $\Sigma x(\log y) = 219.89$, we get
$9.5879 = 6(\log a) + 125.8(\log b)$
$219.89 = 125.8(\log a) + 5{,}338.34(\log b)$
and the solutions of these normal equations are $\log a = 1.4516$, $\log b = 0.00698$, and $a = 28.3$ and $b = 1.02$. Thus, the equation of the exponential curve is $\hat{y} = 28.3(1.02)^x$. Also, for $x = 60$, we get $\log \hat{y} = 1.4516 + 0.00698(60) \approx 1.8704$ and $\hat{y} = 74.2$ mrem/year.

15.59
1. $H_0: \beta = 0.50$ and $H_A: \beta > 0.50$
2. $\alpha = 0.05$
3. Reject the null hypothesis if $t \geq 1.860$.
4. $t = \frac{1.2196 - 0.50}{\frac{0.33}{\sqrt{1.284}}} = 2.47$.
5. The null hypothesis must be rejected.

15.61 We get $S_{yy} = 486{,}017 - \frac{1}{9}(1{,}975)^2 \approx 52{,}614.22$, $s_e = \sqrt{\frac{52{,}614.22 - 7.439(3{,}864.89)}{7}} \approx 58.39$, and $t = \frac{50.001 - 55}{\frac{58.39}{\sqrt{\frac{1}{9} + \frac{22.78^2}{519.56}}}} \approx -0.09$.

1. $H_0: \alpha = 55$ and $H_A: \alpha \neq 55$.
2. $\alpha = 0.01$.
3. Reject the null hypothesis if $t \leq -3.499$ or $t \geq 3.499$.
4. $t = -0.09$.
5. The null hypothesis cannot be rejected.

15.63 (a) The fitted regression equation is
$\hat{y} = -32.702 + 3.819x_1 + 0.058x_2$
where y represents the score, x_1 represents the number of hours, and x_2 represents the SAT score.

(b) $-32.702 + 3.819(12) + 0.058(1,000) \approx 71.13$;
$-32.702 + 3.819(18) + 0.058(1,000) \approx 94.04$;
$-32.702 + 3.819(12) + 0.058(1,200) \approx 82.73$;
$-32.702 + 3.819(18) + 0.058(1,200) \approx 105.64$.

With regard to the final prediction, we should note that the database did not contain anyone with a combination of study time and SAT anywhere near 18 and 1,200.

CHAPTER 16

Correlation

16.1 $r = \dfrac{23.6}{\sqrt{(344)(1.915)}} \approx 0.92$.

16.3 $100\%(-0.01)^2 = 0.01\%$.

16.5 $r = \dfrac{185.875}{\sqrt{(378.875)(196.875)}} \approx 0.68$.

16.7 It should not come as a surprise, since we can always draw a straight line passing through two distinct points. The line which passes through the two points of part (a) has a positive slope and $r = +1$; the line which passes through the two points of part (b) has a negative slope and $r = -1$.

16.9 A correlation is symmetric in the two variables and shows only association (not causation). She should not be asking whether one of the variables influences the other.

16.11
(a) There is a positive correlation.
(b) There is a positive correlation.
(c) There is a positive correlation
(d) There is a no correlation.
(d) There is a negative correlation.

16.13
(a) $r = \dfrac{24{,}780}{\sqrt{(113.2)(233{,}352{,}000)}} \approx 0.15$;

(b) $r = \dfrac{-33}{\sqrt{(113.2)(16)}} \approx -0.78$;

(c) $r = \dfrac{30{,}900}{\sqrt{(16)(233{,}352{,}000)}} \approx 0.51$.

16.15 For instance, if the two correlation coefficients had been 0.80 and 0.90, we would have obtained $100\%(0.80)^2 + 100\%(0.90)^2 = 145\%$, which, of course, is impossible. The conclusion is not valid (that is, the percentages cannot be added) because weight at birth and average daily intake of food are, themselves, related.

16.17 (a) 0.7976; (b) 0.5413.

16.19   1. $H_0: \rho = 0$ and $H_A: \rho \neq 0$.
        2. $\alpha = 0.05$.
        3.   Reject the null hypothesis if $z \leq -1.96$ or $z \geq 1.96$.
        4.   Since $n = 12$ and $r = -0.01$, we get $Z = -0.010$ and $z = -0.010\sqrt{9} = -0.030$.
        5.   The null hypothesis cannot be rejected; the value obtained for r is not significant.

16.21   (a) We get $S_{xx} = 22{,}265 - \frac{1}{12}(507)^2 = 844.25$, $S_{yy} = 1{,}802 - \frac{1}{12}(144)^2 = 74$,

             $S_{xy} = 6{,}314 - \frac{1}{12}(507)(144) = 230$, and $r = \dfrac{230}{\sqrt{(844.25)(74)}} \approx 0.92$.

        (b) 1. $H_0: \rho = 0$ and $H_A: \rho \neq 0$.
             2. $\alpha = 0.05$.
             3.   Reject the null hypothesis if $z \leq -1.96$ or $z \geq 1.96$.
             4.   Since $n = 12$ and $r = 0.92$, we get $Z = 1.589$ and $z = 1.589\sqrt{9} = 4.767$.
             5.   The null hypothesis must be rejected; the value obtained for r is significant.

16.23   (a) 1. $H_0: \rho = 0$ and $H_A: \rho \neq 0$.
             2. $\alpha = 0.05$.
             3.   Reject the null hypothesis if $t \leq -2.228$ or $t \geq 2.228$.
             4. $t = \dfrac{0.50\sqrt{10}}{\sqrt{1-(0.50)^2}} \approx 1.83$.
             5.   The null hypothesis cannot be rejected; the value obtained for r is not significant.

        (b) 1. $H_0: \rho = 0$ and $H_A: \rho \neq 0$.
             2. $\alpha = 0.05$.
             3.   Reject the null hypothesis if $t \leq -2.101$ or $t \geq 2.101$.
             4. $t = \dfrac{0.62\sqrt{18}}{\sqrt{1-(0.62)^2}} \approx 3.35$.
             5.   The null hypothesis must be rejected.

16.25   1. $H_0: \rho = 0$ and $H_A: \rho \neq 0$.
        2. $\alpha = 0.05$.
        3.   Reject the null hypothesis if $t \leq -2.228$ or $t \geq 2.228$.
        4. $t = \dfrac{(-0.01)\sqrt{10}}{\sqrt{1-(-0.01)^2}} \approx -0.03$.
        5.   The null hypothesis cannot be rejected; the value obtained for r is not significant.

16.29   1. $H_0: \rho = 0.75$ and $H_A: \rho \neq 0.75$.
        2. $\alpha = 0.01$.
        3.   Reject the null hypothesis if $z \leq -2.575$ or $z \geq 2.575$.
        4.   Since $n = 13$, $r = 0.94$, and $\rho = 0.75$, we get $Z = 1.738$, $\mu_Z = 0.973$, and $z = (1.738 - 0.973)\sqrt{10} \approx 2.42$.
        5.   The null hypothesis cannot be rejected.

16.31 (a) $1.099 \pm \dfrac{1.96}{\sqrt{12}}$, or 1.099 ± 0.566; $0.533 < \mu_Z < 1.665$, and $0.49 < \rho < 0.93$.

(b) $-0.224 \pm \dfrac{1.96}{\sqrt{27}}$; or -0.224 ± 0.377; $-0.601 < \mu_Z < 0.153$, and $-0.54 < \rho < 0.15$.

(c) $0.758 \pm \dfrac{1.96}{\sqrt{97}}$, or 0.758 ± 0.199; $0.559 < \mu_Z < 0.957$, and $0.51 < \rho < 0.74$.

16.33 Suppose that we use X as the name for the row variable. In the data, we find that 31 students have $X = 1$, 27 students have $X = 2$, 28 students have $X = 3$, and 36 students have $X = 4$. This leads to the totals $\Sigma x = 313$ and $\Sigma x^2 = 967$. With $n = 122$ students, we find $S_{xx} = 967 - \dfrac{313^2}{122} \approx 163.975$. Now let Y be the name of the column variable. There are 4 students with $Y = -3$, 3 students with $Y = -2$, and so on. This gives the sums $\Sigma y = 78$ and $\Sigma y^2 = 236$. Then $S_{yy} = 236 - \dfrac{78^2}{122} \approx 186.131$. We can then find

$\Sigma xy = (1)(-3)(2) + (1)(-2)(1) + (1)(-1)(2) + \cdots + (4)(2)(5) + (4)(3)(4) = 239$. Then

$S_{xy} = 239 - \dfrac{(313)(78)}{122} \approx 38.885$. The correlation can be computed as

$r = \dfrac{38.885}{\sqrt{(163.975)(186.131)}} \approx 0.223$.

This is a modest positive correlation, and we interpret it as saying that the students tend toward greater agreement with the humanities requirement the longer they have attended the college.

16.35 Let X denote the row variable and let Y be the column variable. Use scores (say) $-2, -1, 0, 1$ for the columns. It can be shown that $\Sigma y = -11$ and $\Sigma y^2 = 39$. Then $S_{yy} = 35.3333$. These can be used for both parts (a) and (b).

(a) Use scores $-1, 0,$ and 1 for the three rows. For this choice, $\Sigma x = -1$ and $\Sigma x^2 = 21$. This leads to $S_{xx} = 20.9697$. Also,

$\Sigma xy = 6(2) + 4(1) + 1(0) + 0(-1)$
$\qquad + 0(0) + 2(0) + 7(0) + 3(0)$
$\qquad + 0(-2) + 1(-1) + 4(0) + 5(1) = 20$

leading to $S_{xy} = 20 - (-1)\dfrac{(-11)}{300} \approx 19.9633$. The correlation is then

$r = \dfrac{19.9633}{\sqrt{(20.9697)(35.3333)}} \approx 0.733$.

(b) Use scores 1, 2 an 4 for the three rows. For this choice $\Sigma x = 75$ and $\Sigma x^2 = 219$. This leads to $S_{xx} = 48.5455$. Also,

$$\Sigma xy = 6(-2) + 4(-1) + 1(0) + 0(1)$$
$$+ 0(-4) + 2(-2) + 7(0) + 3(2)$$
$$+ 0(-8) + 1(-4) + 4(0) + 5(4) = 2$$

leading to $S_{xy} = 2 - \dfrac{(75)(-11)}{33} = 27$. The correlation is then

$$r = \dfrac{27}{\sqrt{(48.5455)(35.3333)}} \approx 0.652.$$ For this example, the numeric value of r is little changed when the scoring system is altered a bit.

(c) The score spacings 1, 2, and 4 are equivalent to −1, 0, and 2; the second gap is twice as large as the first gap. The value of r will not be altered.

16.37 $\hat{y} = -16{,}740 + 1{,}961(38) + 5{,}976(4) = 81{,}682$
$\hat{y} = -16{,}740 + 1{,}961(46) + 5{,}976(0) = 73{,}466$
$\hat{y} = -16{,}740 + 1{,}961(39) + 5{,}976(5) = 89{,}619$
$\hat{y} = -16{,}740 + 1{,}961(43) + 5{,}976(?) = 79{,}535$
$\hat{y} = -16{,}740 + 1{,}961(32) + 5{,}976(4) = 69{,}916$.

Since $\bar{y} = \dfrac{394{,}200}{5} = 78{,}840$, we get

$$\Sigma(\hat{y} - \bar{y})^2 = (81{,}682 - 78{,}840)^2 + (73{,}466 - 78{,}840)^2 + (89{,}619 - 78{,}840)^2$$
$$+ (79{,}535 - 78{,}840)^2 + (69{,}916 - 78{,}840)^2 = 233{,}264{,}482$$

$\Sigma(y - \bar{y})^2 = 31{,}312{,}080{,}000 - \dfrac{1}{5}(394{,}200)^2 = 233{,}352{,}000$, and $\sqrt{\dfrac{233{,}264{,}482}{233{,}352{,}000}} \approx 0.9998$.

The correlation between income and age was 0.15, while the correlation between income and years of college was 0.51. It can be seen that R is much larger than either of these.

16.39 The multiple correlation coefficient based on both variables cannot be numerically less than the ordinary correlation coefficient based on only one of the two variables.

16.41 (a) $r_{13.2} = \dfrac{0.15 - (-0.78)(0.51)}{\sqrt{1 - (-0.78)^2}\sqrt{1 - (0.51)^2}} \approx 1.00;$

(b) $r_{23.1} = \dfrac{0.51 - (-0.78)(0.15)}{\sqrt{1 - (-0.78)^2}\sqrt{1 - (0.15)^2}} \approx 1.00.$

16.43 No. When $r = 0.90$ then $100(0.90)^2 = 81$ percent of the variation of the y's is accounted for by the relationship of x; and when $r = 0.45$ then $100(0.45)^2 = 20.25$ of the variation of the y's is accounted for by the relationship of x. We can then say that the 0.90 correlation is $\frac{81}{20.25} = 4$ times as strong as a relationship for which $r = 0.45$.

16.47 (a)
1. $H_0: \rho = 0$ and $H_A: \rho \neq 0$.
2. $\alpha = 0.05$.
3. Reject the null hypothesis if $|z| \geq 1.96$, where $z = Z \cdot \sqrt{n-3}$. Otherwise accept the null hypothesis or reserve judgment.
4. Table IX indicates that $r = 0.36$ corresponds to $Z = 0.377$; along with $n = 22$, we find $z = 0.377\sqrt{19} \approx 1.64$.
5. Since $1.64 < 1.96$, we cannot reject the null hypothesis.

(b)
1. $H_0: \rho = 0$ versus $H_A: \rho \neq 0$.
2. $\alpha = 0.05$.
3. Reject the null hypothesis if $|t| \geq 2.086$, where $t = \frac{r\sqrt{n-2}}{\sqrt{1-r^2}}$ and 2.086 is $t_{0.025}$ for $n - 2 = 20$ degrees of freedom. Otherwise accept the null hypothesis or reserve judgment.
4. Using $r = 0.36$ and $n = 22$, we find $t = 1.73$.
5. Since $1.73 < 2.086$, we canot reject the null hypothesis.

16.49
1. $H_0: \rho = 0$ and $H_A: \rho \neq 0$.
2. $\alpha = 0.05$.
3. Reject the null hypothesis if $z \leq -1.96$ or $z \geq 1.96$.
4. Since $n = 6$ and $r = -0.92$, we get $Z = -1.589$ and $z = -1.589\sqrt{3} \approx -2.75$.
5. The null hypothesis must be rejected; the value obtained for r is significant.

16.51 (a) $0.472 \pm \frac{1.96}{\sqrt{17}}$, or 0.472 ± 0.475; $-0.003 < \mu_Z < 0.947$ and $0 < \rho < 0.74$;

(b) $-0.332 \pm \frac{1.96}{\sqrt{35}}$, or -0.332 ± 0.331; $-0.663 < \mu_Z < -0.001$ and $-0.58 < \rho < 0$.

16.53 $\sqrt{1 - \frac{926}{1,702}} \approx 0.68$.

16.55 $r = \frac{2,111.25}{\sqrt{(2,068.875)(2,907.5)}} \approx 0.86$.

16.57 Since $n = 8$, $\Sigma x = 49$, $\Sigma x^2 = 2,369$, $\Sigma y = 38$, $\Sigma y^2 = 3,008$, and $\Sigma xy = 2,344$, we get $S_{xx} = 2,368 - \frac{1}{8}(49)^2 = 2,068.875$, $S_{yy} = 3,088 - \frac{1}{8}(38)^2 = 2,907.5$, and $S_{xy} = 2,344 - \frac{1}{8}(49)(38) = 2,111.25$. The values of S_{xx}, S_{yy}, and S_{xy} are the same as in Exercise 16.56, so the value of r is again 0.86.

CHAPTER 17

Nonparametric Tests

17.1
1. $H_0: \tilde{\mu} = 6.0$ and $H_A: \tilde{\mu} \neq 6.0$.
2. $\alpha = 0.05$.
3'. Reject the null hypothesis if the probability of x or fewer plus signs, or that of x or more plus signs, is less than or equal to 0.025.
4'. Getting $-+++-++++--+-+$ and $x = 9$, we find that the probability of $x \leq 9$ is 0.909 and the probability of $x \geq 9$ is 0.212.
5'. The null hypothesis cannot be rejected; there is no real evidence that the median time he has to wait is not 6.0 minutes.

17.3
1. $H_0: \tilde{\mu} = 100.0$ and $H_A: \tilde{\mu} \neq 100.0$.
2. $\alpha = 0.05$.
3'. Reject the null hypothesis if the probability of x or fewer plus signs, or that of x or more plus signs, is less than or equal to 0.05.
4'. Getting $+++-+++-++-+$ (where two values are discarded) and $x = 9$, we find that the probability of $x \leq 9$ is 0.981 and the probability of $x \geq 9$ is 0.073.
5'. The null hypothesis cannot be rejected; there is no real evidence that the median weight of the packages is not 100.0 grams.

17.5
1. $H_0: \mu_1 = \mu_2$ and $H_A: \mu_1 > \mu_2$; could also have $H_0: \tilde{\mu}_D = 0$ and $H_A: \tilde{\mu}_D > 0$.
2. $\alpha = 0.01$.
3'. Reject the null hypothesis if the probability of x or more plus signs is less than or equal to 0.01.
4'. Getting $++-++++$ and $x = 6$, we find from Table V that for $n = 7$ and $p = 0.50$ the probability of $x \geq 6$ is 0.063.
5'. The null hypothesis cannot be rejected; there is no real evidence that on the average the first flight carries more passengers than the second flight.

17.7
1. $H_0: \mu_1 = \mu_2$ and $H_A: \mu_1 \neq \mu_2$; could also have $H_0: \tilde{\mu}_D = 0$ and $H_A: \tilde{\mu}_D \neq 0$.
2. $\alpha = 0.01$.
3'. Reject the null hypothesis if the probability of x or more plus signs, or that of x or fewer plus signs, is less than or equal to 0.005.
4'. Getting $----+--+-+$ and $x = 3$, we find that the probability of $x \leq 3$ is 0.172 and the probability of $x \geq 3$ is 0.945.
5'. The null hypothesis cannot be rejected; there is no real evidence of a systematic difference.

17.9 1. $H_0: \tilde{\mu} = 100.0$ and $H_A: \tilde{\mu} \neq 100.0$.
2. $\alpha = 0.01$.
3. Reject the null hypothesis if $z \leq -2.575$ or $z \geq 2.575$.
4. Getting $+++-+++-++-+$ (where two values are discarded) and $x = 9$, we find that
$$z = \frac{9 - 12(0.50)}{\sqrt{12(0.50)(0.50)}} \approx 1.73.$$
5. The null hypothesis cannot be rejected; there is no real evidence that the median weight of the packages is not 100.0 grams.

17.11 1. $H_0: \tilde{\mu}_D = 0$ and $H_A: \tilde{\mu}_D \leq 0$.
2. $\alpha = 0.05$.
3. Reject the null hypothesis if $z \leq -1.645$.
4. Getting $--+-----+--+------+---+--$ (where data for day 2 is discarded) and $x = 5$, we find that $z = \dfrac{5 - 24(0.50)}{\sqrt{24(0.50)(0.50)}} \approx -2.86$.
5. The null hypothesis must be rejected.

17.15 (a) We use the statistic T and reject the null hypothesis if $T \leq 8$.
(b) We use the statistic T^- and reject the null hypothesis if $T^- \leq 11$.
(c) We use the statistic T^+ and reject the null hypothesis if $T^+ \leq 11$.

17.17 (a) We use the statistic T^- and reject the null hypothesis if $T^- \leq 10$.
(b) We use the statistic T and reject the null hypothesis if $T \leq 7$.
(c) We use the statistic T^+ and reject the null hypothesis if $T^+ \leq 10$

17.19 (a) 1. $H_0: \tilde{\mu} = 40$ and $H_A: \tilde{\mu} < 40$.
2. $\alpha = 0.05$.
3. Reject the null hypothesis if $T^+ \leq 21$.
4. $T^+ = 1 + 11 + 6 = 18$.
5. The null hypothesis must be rejected; on the average the newspaper lists fewer than 40 apartments for rent.
(b) 1. $H_0: \tilde{\mu} = 40$ and $H_A: \tilde{\mu} \neq 40$.
2. $\alpha = 0.05$.
3. Reject the null hypothesis if $T \leq 17$.
4. $T^+ = 18$, $T^- = \dfrac{13 \cdot 14}{2} - 18 = 73$, so that $T = 18$.
5. The null hypothesis cannot be rejected; there is no real evidence that on the average the newspaper does not list 40 apartments for rent.

17.21 1. $H_0: \tilde{\mu} = 100.0$ and $H_A: \tilde{\mu} \neq 100.0$.
2. $\alpha = 0.01$.
3. Reject the null hypothesis if $T \leq 7$.
4. $T^- = 8 + 11 + 1 = 20$, $T^+ = \dfrac{12 \cdot 13}{2} - 20 = 58$, and $T = 20$.
5. The null hypothesis cannot be rejected; there is no real evidence that the mean weight of the packages is not 100.0 grams.

17.23 1. $H_0: \tilde{\mu} = 35$ and $H_A: \tilde{\mu} \neq 35$.
2. $\alpha = 0.05$.
3. Reject the null hypothesis if $T \leq 11$.
4. The differences from 3.5, with the absolute ranks below, are these:
    −0.5 2.0 −0.1 2.7 −0.6 5.5 1.7 0.3 −1.1 9.0 4.5
      3   7   1   8   4   10 6 2   5   11 9
    Then $T^+ = 7+8+10+6+2+11+9 = 53$ and $T^- = 3+1+4+5 = 13$, so that $T = 13$.
5. Since $T > 11$, the null hypothesis cannot be rejected.

17.25 (a) 1. $H_0: \tilde{\mu}_1 = \tilde{\mu}_2$ and $H_A: \tilde{\mu}_1 > \tilde{\mu}_2$; could use $H_0: \tilde{\mu}_D = 0$ and $H_A: \tilde{\mu}_D > 0$.
2. $\alpha = 0.05$.
3. Reject the null hypothesis if $T^- \leq 36$.
4. $T^- = 2 + 9.5 + 14 = 25.5$.
5. The null hypothesis must be rejected.

(b) 1. $H_0: \tilde{\mu}_1 = \tilde{\mu}_2$ and $H_A: \tilde{\mu}_1 > \tilde{\mu}_2$; could use $H_0: \tilde{\mu}_D = 0$ and $H_A: \tilde{\mu}_D > 0$.
2. $\alpha = 0.05$.
3. Reject the null hypothesis if $z \geq 1.645$.
4. $\mu_{T^+} = \frac{16 \cdot 17}{4} = 68$, $\sigma_{T^+} = \sqrt{\frac{16 \cdot 17 \cdot 33}{24}} \approx 19.34$ so that $z = \frac{110.5 - 68}{19.34} \approx 2.20$.
5. The null hypothesis must be rejected; the bank averages more 3-month certificates.

17.27 (a) 1. $H_0: \tilde{\mu} = 37.0$ and $H_A: \tilde{\mu} \neq 37.0$.
2. $\alpha = 0.05$.
3. Reject the null hypothesis if $T \leq 81$.
4. $T^- = 11 + 15 + 4 + 2 + 1 + 6 + 3 = 42$, $T^+ = \frac{24 \cdot 25}{2} - 42 = 258$ and $T = 42$.
5. The null hypothesis must be rejected; the mean weight of the suitcases is not 37.0 pounds.

(b) 1. $H_0: \tilde{\mu} = 37.0$ and $H_A: \tilde{\mu} \neq 37.0$.
2. $\alpha = 0.05$.
3. Reject the null hypothesis if $z \leq -1.96$ or $z \geq 1.96$.
4. $\mu_{T^+} = \frac{24 \cdot 25}{4} = 150$, and $\sigma_{T^+} = \sqrt{\frac{24 \cdot 25 \cdot 49}{24}} = 35$ so that $z = \frac{258 - 150}{35} \approx 3.09$.
5. The null hypothesis must be rejected; the mean weight of the suitcases is not 37.0 pounds.

17.29 (a) We use the statistic U_2 and reject the null hypothesis if $U_2 \leq 21$;
(b) We use the statistic U and reject the null hypothesis if $U \leq 17$;
(c) We use the statistic U_1 and reject the null hypothesis if $U_1 \leq 21$;
(d) We use the statistic U_2 and reject the null hypothesis if $U_2 \leq 14$;
(e) We use the statistic U and reject the null hypothesis if $U \leq 11$;
(f) We use the statistic U_1 and reject the null hypothesis if $U_1 \leq 14$.

17.31 We use the statistic U and reject the null hypothesis if
(a) $U \leq 2$; (b) $U \leq 15$; (c) $U \leq 11$; (d) $U \leq 1$.

17.33 For $n_1 = 3$ and $n_2 = 3$ the probability of $U \leq 0$ is 0.10, which already exceeds 0.05.

114 Chapter 17 Nonparametric Tests

17.35 1. $H_0: \mu_1 = \mu_2$ and $H_A: \mu_1 \neq \mu_2$.
2. $\alpha = 0.05$.
3. Reject the null hypothesis if $U \leq 49$.
4. $W_1 = 18 + 2 + 9 + 10 + 5 + 16 + 27 + 11 + 9 + 20 + 14 + 23 + 6 + 25 + 23 + 3 = 208$, so that $U_1 = 208 - \frac{15 \cdot 16}{2} = 88$, $U_2 = 15 \cdot 12 - 88 = 92$, and $U = 88$.
5. The null hypothesis must be rejected; men and women do not take equally long to complete the test.

17.37 1. $H_0: \mu_1 = \mu_2$ and $H_A: \mu_1 < \mu_2$.
2. $\alpha = 0.05$.
3. Reject the null hypothesis if $U_1 \leq 10$.
4. $W_1 = 8 + 1 + 3.5 + 5 + 2 + 7 = 26.5$, so that $U_1 = 26.5 - \frac{6 \cdot 7}{2} = 5.5$
5. The null hypothesis must be rejected; the castings of production lot B are on the average harder than those of production lot A.

17.39 1. $H_0: \mu_1 = \mu_2$ and $H_A: \mu_1 \neq \mu_2$.
2. $\alpha = 0.05$.
3. Reject the null hypothesis if $z \leq -1.96$ or $z \geq 1.96$.
4. Since $\mu_{U_1} = \frac{9 \cdot 10}{2} = 45$ and $\mu_{U_1} = \sqrt{\frac{9 \cdot 10 \cdot 20}{12}} \approx 12.25$, and we found in the text that $U_1 = 24$, we get $z = \frac{24 - 45}{12.25} \approx -1.71$.
5. The null hypothesis cannot be rejected; there is no real evidence that there is a difference in the average grain size.

17.41 1. $H_0: \mu_1 = \mu_2$ and $H_A: \mu_1 \neq \mu_2$.
2. $\alpha = 0.05$.
3. Reject the null hypothesis if $z \leq -1.96$ or $z \geq 1.96$.
4. Since $W_1 = 499$, $U_1 = 499 - \frac{20 \cdot 21}{2} = 289$, $\mu_{U_1} = \frac{20 \cdot 20}{2} = 200$, and $\sigma_{U_1} = \sqrt{\frac{20 \cdot 20 \cdot 41}{12}} \approx 36.97$, we get $z = \frac{289 - 200}{36.97} \approx 2.41$.
5. The null hypothesis must be rejected; there is a difference in the average knowledge of American history between freshmen entering the two universities.

17.43 1. $H_0: \mu_1 = \mu_2$ and $H_A: \mu_1 \neq \mu_2$.
2. $\alpha = 0.05$.
3. Reject the null hypothesis if $z \leq -1.96$ or $z \geq 1.96$.
4. Since $U_1 = 110$, $\mu_{U_1} = \frac{12 \cdot 12}{2} = 72$, and $\sigma_{U_1} = \sqrt{\frac{12 \cdot 12 \cdot 25}{12}} \approx 17.32$, we get $z = \frac{110 - 72}{17.32} \approx 2.19$.
5. The null hypothesis must be rejected; students from the two groups cannot be expected to score equally well on the test.

Modern Elementary Statistics – 9th Edition **115**

17.45 1. $H_0: \mu_1 = \mu_2 = \mu_3$ and $H_A: \mu_1, \mu_2,$ and μ_3 are not all equal.
2. $\alpha = 0.05$.
3. Reject the null hypothesis if $\chi^2 \geq 5.991$.
4. Since $R_1 = 14 + 1 + 16 + 11 + 9 + 13 = 64$, $R_2 = 2 + 12 + 5 + 18 + 16 + 6.5 = 59.5$, and $R_3 = 4 + 8 + 16 + 6.5 + 3 + 10 = 47.5$, we get
$$H = \frac{12}{18 \cdot 19}\left(\frac{64^2}{6} + \frac{59.5^2}{6} + \frac{47.5^2}{6}\right) - 3 \cdot 19 \approx 0.85.$$
5. The null hypothesis cannot be rejected; there is no real evidence that the average mileage yields are not all the same.

17.47 1. $H_0: \mu_1 = \mu_2 = \mu_3 = \mu_4$ and $H_A: \mu_1, \mu_2, \mu_3,$ and μ_4 are not all equal.
2. $\alpha = 0.05$.
3. Reject the null hypothesis if $\chi^2 > 7.815$.
4. $R_1 = 5.5 + 15.0 + 3.5 + 8.5 + 12.0 = 44.5$,
$R_2 = 8.5 + 15.0 + 2.0 + 20.0 + 12.0 = 57.5$,
$R_3 = 5.5 + 18.0 + 15.0 + 8.5 + 18.0 = 65.0$,
$R_4 = 18.0 + 1.0 + 8.5 + 12.0 + 3.5 = 43.0$, and we get
$$H = \frac{12}{20(20+1)}\left(\frac{44.5^2}{5} + \frac{57.5^2}{5} + \frac{65.0^2}{5} + \frac{43.0^2}{5}\right) - 3 \cdot 21 \approx 1.9.$$
5. The null hypothesis cannot be rejected.

17.49 1. $H_0: \mu_1 = \mu_2 = \mu_3 = \mu_4$ and $H_A: \mu_1, \mu_2, \mu_3,$ and μ_4 are not all equal.
2. $\alpha = 0.05$.
3. Reject the null hypothesis if $\chi^2 \geq 7.815$.
4. Since $R_1 = 4 + 7 + 10 + 14 + 18 = 53$, $R_2 = 5 + 12 + 15 + 16 + 20 = 68$, $R_3 = 1 + 3 + 6 + 9 + 11 = 30$, and $R_4 = 2 + 8 + 13 + 17 + 19 = 59$ we get
$$H = \frac{12}{20 \cdot 21}\left(\frac{53^2}{5} + \frac{68^2}{5} + \frac{30^2}{5} + \frac{59^2}{5}\right) - 3 \cdot 21 \approx 4.51.$$
5. The null hypothesis cannot be rejected; there is no real evidence that on the average the bowler does not perform equally well with the four balls.

17.51 1. H_0: Arrangement is random.
H_A: Arrangement is not random.
2. $\alpha = 0.05$.
3. Reject the null hypothesis if $u \leq 6$ or $u \geq 16$.
4. $n_1 = 14$, $n_2 = 8$ and $u = 5$.
5. The null hypothesis must be rejected; the arrangement is not random.

17.53 1. H_0: Arrangement is random.
H_A: Arrangement is not random.
2. $\alpha = 0.01$.
3. Reject the null hypothesis if $u \leq 4$.
4. $n_1 = 14$, $n_2 = 6$ and $u = 5$.
5. The null hypothesis cannot be rejected; there is no real evidence of any lack of randomness.

17.55 1. H_0: Arrangement is random.
   H_A: Arrangement is not random.
2. $\alpha = 0.05$.
3. Reject the null hypothesis if $z \leq -1.96$ or $z \geq 1.96$.
4. Since $n_1 = 28$, $n_2 = 22$, and $u = 23$, we get $\mu_u = \dfrac{2 \cdot 28 \cdot 22}{50} + 1 = 25.64$, $\sigma_u = \sqrt{\dfrac{2 \cdot 28 \cdot 22 \cdot 1{,}182}{50^2 \cdot 49}} \approx 3.45$, and $z = \dfrac{23 - 25.64}{3.45} \approx -0.77$.
5. The null hypothesis cannot be rejected; there is no real evidence of any lack of randomness.

17.57 Various

17.59 1. H_0: Arrangement is random.
   H_A: Arrangement is not random.
2. $\alpha = 0.05$.
3. Reject the null hypothesis if $z \leq -1.96$ or $z \geq 1.96$.
4. Since $n_1 = 26$, $n_2 = 24$, and $u = 26$, we get $\mu_u = \dfrac{2 \cdot 26 \cdot 24}{50} + 1 = 25.96$, $\sigma_u = \sqrt{\dfrac{2 \cdot 26 \cdot 24 \cdot 1{,}198}{50^2 \cdot 49}} \approx 3.49$, and $z = \dfrac{26 - 25.95}{3.49} \approx 0.01$.
5. The null hypothesis cannot be rejected; there is no real evidence that the arrangement is not random.

17.61 1. H_0: Arrangement is random.
   H_A: Arrangement is not random.
2. $\alpha = 0.01$.
3. Reject the null hypothesis if $u \leq 7$ or $u \geq 22$.
4. Since the median is 5, the arrangement of values above and below the median is aaaaa aaabb bbbbb aaaaa bbbbb ba and $n_1 = 14$, $n_2 = 13$, and $u = 5$.
5. The null hypothesis must be rejected; the arrangement is not random.

17.63 1. H_0: Arrangement is random.
   H_A: There is a trend.
2. $\alpha = 0.05$.
3. Reject the null hypothesis if $z \leq -1.645$.
4. Since the median is 66, the arrangement of values above and below the median is aaaaa aaaaa bbaaa abbaa bbbaa bbbbb bbbab bbbab and $n_1 = 20$, $n_2 = 20$, and $u = 12$. Thus, $\mu_u = \dfrac{2 \cdot 20 \cdot 20}{40} + 1 = 21$, $\sigma_u = \sqrt{\dfrac{2 \cdot 20 \cdot 20 \cdot 760}{40^2 \cdot 39}} \approx 3.12$, and $z = \dfrac{12 - 21}{3.12} \approx -2.88$.
5. The null hypothesis must be rejected; there is a real trend.

17.65

Rank of x	Rank of y	d	d^2
10	10	0	0
6	5	1	1
8	9	−1	1
2	3	−1	1
12	12	0	0
5	4	1	1
9	8	1	1
3	7	−4	16
11	11	0	0
4	2	2	4
7	6	1	1
1	1	0	0
			26

$r_S = 1 - \dfrac{6 \cdot 26}{12 \cdot 143} \approx 0.91$

17.67
1. $H_0: \rho = 0$ and $H_A: \rho \neq 0$.
2. $\alpha = 0.05$.
3. Reject the null hypothesis if $z \leq -1.96$ or $z \geq 1.96$.
4. $z = 0.65\sqrt{11} \approx 2.16$.
5. The null hypothesis must be rejected; there is a relationship between the amount of time it takes a mechanic to assemble the piece of machinery in the morning and in the late afternoon.

17.69

Rank of x	Rank of y	d	d^2
4	7.5	−3.5	12.25
9	6	3	9
3	4	−1	1
1	2	−1	1
5	3	2	4
8	9	−1	1
6	7.5	−1.5	2.25
2	1	1	1
7	5	2	4
			35.50

$r_S = 1 - \dfrac{6(35.5)}{9 \cdot 80} \approx 0.70$

1. $H_0: \rho = 0$ and $H_A: \rho \neq 0$.
2. $\alpha = 0.01$.
3. Reject the null hypothesis if $z \leq -2.575$ or $z \geq 2.575$.
4. $z = 0.70\sqrt{8} \approx 1.98$.
5. The null hypothesis cannot be rejected; the value obtained for r_S is not significant.

17.71 1. $H_0: \rho = 0$ and $H_A: \rho \neq 0$.
2. $\alpha = 0.01$.
3. Reject the null hypothesis if $z \leq -2.575$ or $z \geq 2.575$.
4. Since $n = 40$ and $r_S = 0.48$, we get $z = 0.48\sqrt{39} \approx 3.00$.
5. The null hypothesis must be rejected; the value obtained for r_S is significant.

17.73 Since $r_S = 1 - \dfrac{6 \cdot 64}{10(10^2 - 1)} \approx 0.61$ for judges A and B, $r_S = 1 - \dfrac{6 \cdot 714}{990} \approx -0.05$ for judges A and C,

and $r_S = 1 - \dfrac{6 \cdot 194}{990} \approx -0.18$ for judges B and C, we find that

(a) the rankings of judges A and B are most alike;
(b) the rankings of judges B and C differ the most.

17.75 1. $H_0: \tilde{\mu} = 55.00$ and $H_A: \tilde{\mu} \neq 55.00$.
2. $\alpha = 0.05$.
3. Reject the null hypothesis if $T \leq 21$.
4. Since $T^+ = 10 + 13 + 7 + 9 + 2 + 11 + 12 + 8 + 14 = 86$ and $T^- = 5 + 1 + 3 + 4 + 6 = 19$, we get $T = 19$.
5. The null hypothesis must be rejected; the mean price is not $55.00.

17.77 1. H_0: Arrangement is random.
   H_A: Arrangement is not random.
2. $\alpha = 0.05$.
3. Reject the null hypothesis if $u \leq 4$.
4. $n_1 = 15$, $n_2 = 5$, and $u = 4$.
5. The null hypothesis must be rejected; the arrangement is not random.

17.79 1. $H_0: \mu_1 = \mu_2$ and $H_A: \mu_1 \neq \mu_2$.
2. $\alpha = 0.05$.
3. Reject the null hypothesis if $U \leq 37$.
4. Since $W_1 = 12 + 24 + 19.5 + 17 + 13.5 + 18 + 8 + 16 + 4 + 22 + 19.5 + 13.5 = 187$, we get $U_1 = 187 - \dfrac{12 \cdot 13}{2} = 109$, $U_2 = 144 - 109 = 35$, and $U = 35$.
5. The null hypothesis must be rejected; on the average the two insecticides are not equally effective.

17.81

Rank of x	Rank of y	d	d^2
5	12	-7	49
13	7	6	36
11	5	6	36
12	14	-2	4
10	2.5	7.5	56.25
1	2.5	-1.5	2.25
9	15	-6	36
7.5	7	0.5	0.25
14	10	4	16
6	11	-5	25
3	1	2	4
2	4	-2	4
7.5	13	-5.5	30.25
4	7	-3	9
15	9	6	36
			344

$$r_S = 1 - \frac{6 \cdot 344}{15(15^2 - 1)} \approx 0.39$$

1. H_0: $\rho = 0$ and H_A: $\rho \neq 0$.
2. $\alpha = 0.01$.
3. Reject the null hypothesis if $z \leq -2.575$ or $z \geq 2.575$.
4. Since $n = 15$ and $r_S = 0.39$, we get $z = 0.39\sqrt{14} \approx 1.46$.
5. The null hypothesis cannot be rejected; the value obtained for r_S is not significant.

17.83
1. H_0: $\tilde{\mu} = 169$ and H_A: $\tilde{\mu} \neq 169$.
2. $\alpha = 0.05$.
3. Reject the null hypothesis if $T \leq 11$.
4. Since $T^+ = 1+7+11+6+3+10+5+8+4 = 55$ and $T^- = 9+2 = 11$, we get $T = 11$.
5. The null hypothesis must be rejected; the population median is not 169.

17.85
1. $H_0: \mu_1 = \mu_2$ and $H_A: \mu_1 \neq \mu_2$; also $H_0: \tilde{\mu}_D = 0$ and $H_A: \tilde{\mu}_D \neq 0$.
2. $\alpha = 0.05$.
3. Reject the null hypothesis if $z \leq -1.96$ or $z \geq 1.96$.
4. Getting + + + - - + + + - - + + - + - + + + - + and $x = 13$, we find that
$$z = \frac{13 - 20(0.50)}{\sqrt{20(0.50)(0.50)}} \approx 1.34.$$
5. The null hypothesis cannot be rejected; there is no real evidence that on the average there are not equally many burglaries per day in the two cities.

17.87
(a) We use the statistic U_1 and reject the null hypothesis if $U_1 \leq 23$.
(b) We use the statistic U_2 and reject the null hypothesis if $U_2 \leq 23$.
(c) We use the statistic U and reject the null hypothesis if $U \leq 19$.

17.89 1. $H_0: \rho = 0$ and $H_A: \rho \neq 0$.
2. $\alpha = 0.05$.
3. Reject the null hypothesis if $z \leq -1.96$ or $z \geq 1.96$.
4. Since $n = 25$ and $r_S = 0.27$, we get $z = 0.27\sqrt{24} \approx 1.32$.
5. The null hypothesis cannot be rejected; the value obtained for r_S is not significant.

17.91 1. H_0: Arrangement is random.
   H_A: Arrangement is not random.
2. $\alpha = 0.01$.
3. Reject the null hypothesis if $u \leq 4$ or $u \geq 16$.
4. Since the median is 375, we get aaaaa abbbb baaab bbb (where two values are discarded), $n_1 = 9$, $n_2 = 9$, and $u = 4$.
5. The null hypothesis must be rejected; the arrangement is not random.

17.93 1. $H_0: \mu_1 = \mu_2$ and $H_A: \mu_1 > \mu_2$.
2. $\alpha = 0.05$.
3. Reject the null hypothesis if $U_2 \leq 17$.
4. Since $W_2 = 15 + 5 + 9 + 1 + 6 + 3.5 + 7 = 46.5$, we get $U_2 = 46.5 - \dfrac{7 \cdot 8}{2} = 18.5$.
5. The null hypothesis cannot be rejected; there is no real evidence that on the average the first ambulance service is slower than the second.

17.95 (a) We use the statistic T^+ and reject the null hypothesis if $T^+ \leq 14$.
(b) We use the statistic T^- and reject the null hypothesis if $T^- \leq 14$.
(c) We use the statistic T and reject the null hypothesis if $T \leq 11$.